GREEN CARBON MATERIALS

GREEN CARBON MATERIALS

ADVANCES AND APPLICATIONS

edited by
Thomas E. Rufford
Denisa Hulicova-Jurcakova
John Zhu

PAN STANFORD PUBLISHING

Published by

Pan Stanford Publishing Pte. Ltd.
Penthouse Level, Suntec Tower 3
8 Temasek Boulevard
Singapore 038988

Email: editorial@panstanford.com
Web: www.panstanford.com

British Library Cataloguing-in-Publication Data
A catalogue record for this book is available from the British Library.

Green Carbon Materials: Advances and Applications

ISBN 978-981-4411-13-4 (Hardcover)
ISBN 978-981-4411-14-1 (eBook)

Printed in the USA

Contents

Preface

Carbon is a versatile material that forms four known allotropes (amorphous, graphite, diamond, and fullerenes) and can be produced in a wide range of forms, including powders, foams, monoliths, cloth fibers, thin films, and granular particles. The physicochemical properties of carbon materials such as surface area, pore size, surface chemistry, and electrical conductivity can be tuned through the choice of precursor materials, preparation methods, and various secondary treatments. The versatility and possibilities to control the properties of carbon have seen carbon materials such as activated carbons used in a wide range of applications, including as adsorbents for wastewater treatment and gas purification, as energy storage materials, and as catalysts.

Novel carbon nanomaterials such as carbon nanotubes have shown great potential as electronic materials, catalysts, and adsorbents. However, significant advances in the carbon nanomaterial production techniques are required if the cost of materials such as carbon nanotubes are to be reduced and their available volumes are to be increased to a level that would allow their widespread industrial application. In contrast, activated carbons, graphite, and graphitic materials have been used for many years as economically viable adsorbents, catalyst supports, and catalyst materials. In this book, we examine the recent advances in technologies to produce low-cost carbons using less energy-intensive production processes and the application of these so-called "green carbons" in the development of cleaner and more efficient energy production and utilization processes. Of particular interest in this book is the use of agricultural and food industry waste materials as a feedstock for activated carbon production.

The content of this book is organized into 10 chapters written by experts in their fields. The first chapter provides an introduction to the functional groups that can be found on the surface of carbon materials described in the rest of the book, including a brief description of the most commonly used analytical techniques to characterize these surface functional groups. Chapter 2 provides an overview of the technologies used to produce activated carbon from lignocellulosic biomass materials, including conventional methods such as chemical and thermal activation and novel methods such as hydrothermal carbonization.

Chapters 3 to 5 highlight the recent advances in the use of carbon materials in energy storage and conversion. Carbons with high specific surface areas have been studied for the adsorption of hydrogen and methane for use onboard hydrogen- or natural gas–powered vehicles (Chapter 3, by Alcañiz-Monge *et al.*) as well as electrode materials for electric double-layer capacitors (Chapter 4, by Rufford *et al.*) and lithium-ion batteries (Chapter 5, by Rooke *et al.*). Chapter 5 also covers the use of cellulose-based carbons as platinum catalyst supports in proton exchange membrane fuel cells (PEMFC).

The second half of the book (Chapters 6 to 10) focuses on the use of carbon materials in pollution control, including the capture of toxic pollutants and volatile organic compounds (VOCs) from wastewater and industrial gases. Chapter 6 (Matos) reports the effect of the surface chemistry of activated carbon prepared from saw dust on the degradation of 4-chlorophenol by photooxidation in the presence of TiO_2. Chapter 7 (Bandosz) surveys the recent advances in the use of activated carbons to remove toxic gases such as ammonia, hydrogen sulfide, and sulfur dioxide from the emissions of industrial processes. This chapter studies the role of surface functional groups, the inorganic content of the carbon, and carbon pore structure in adsorption-reactive adsorption mechanisms involving the toxic gases. Chapter 8 (Cannon and Nieto-Delgado) presents the tailoring of the activated carbon properties for removing the natural organic matter, algal odorants, and endocrine disrupting compounds from water, along with case studies from municipal wastewater treatment plants. In Chapter 9, Dr. Peter Lodewyckx reviews the fundamentals of the adsorption of VOCs on carbon and the kinetics of the adsorption of VOCs in fixed beds. The final chapter (Chapter 10, by Jana and Gamesan) investigates the use of carbon foams prepared from sucrose as a radionuclide trap in sodium-cooled fast nuclear reactors.

The multiple authors in this book have covered many of the significant industrial applications of carbon materials. However, the content of the book is not exhaustive and some notable omissions from the scope of the book include the use of carbon materials as catalysts and catalyst supports (aside from the discussion of PEMFC catalysts in Chapter 5) and the capture of carbon dioxide from flue gases. Each of these topics is significant industrially and scientifically, and as such have been covered in detail in other books (for example, *Carbon Materials for Catalysis*, edited by Phillipe Serp and José Luis Figueiredo, Wiley 2008) and review articles (for example, Marta Sevilla and Antonio B. Fuetes in *Energy & Environmental Science*, 2011, **4**, pp. 1765–1771).

We thank all the authors for their contributions and efforts to submit materials in a timely manner. We also thank our colleagues at the University of Western Australia and the University of Queensland who reviewed sections of this book and provided helpful suggestions. We also wish to thank Stanford Chong and Sarabjeet Garcha at Pan Stanford for their helpful advice and support (including their reminders of deadlines) during various stages in the preparation of this manuscript.

Thomas E. Rufford
Denisa Hulicova-Jurcakova
Zhonghua (John) Zhu

Chapter 1

Surface Chemistry of Green Carbons

Conchi O. Ania

Instituto Nacional del Carbón (INCAR-CSIC), 33011 Oviedo, Asturias, Spain

conchi.ania@incar.csic.es

1.1 Introduction

Recent advances in the design of carbon materials have contributed to the development of highly featured carbons that offer unexpected opportunities in multidisciplinary fields beyond traditional uses [1–4]. Owing to an unforeseen control in the physicochemical and structural properties (particle size and shapes, uniformity in porous void, flexibility) during the synthesis, modern carbons are emerging as promising materials in many science and engineering fields [2–4]. The majority of these applications derive from the flexible coordination chemistry of carbon atoms and their ability to bind other heteroatoms either on the surface or inserted within the structural framework.

Green Carbon Materials: Advances and Applications
Edited by Thomas E. Rufford, Denisa Hulicova-Jurcakova, John Zhu

ISBN 978-981-4411-13-4 (Hardcover), 978-981-4411-14-1 (eBook)
www.panstanford.com

The versatile surface chemistry of carbons provides a useful and unlimited tool to design materials to face new environmental challenges. Small amounts of heteroatoms condition the physico-chemical properties of carbon materials (self-organization, chemical stability, reactivity) [3, 4]; thus the performance of a given carbon material in a specific application can be modulated by selecting an adequate precursor and controlling the structure and chemical composition.

Nowadays, tailoring the features of carbon materials for optimizing, for instance, the adsorption capacity and selectivity of a carbon adsorbent, or the selectivity and efficiency of a catalyst, is not enough: It also becomes essential to balance the environmental and economic performance. The idea is rather simple, as it relies on reducing the environmental impacts associated with the materials' processing and implementation without sacrificing the performance.

In this regard, the use renewable resources and wastes as precursors represents an important strategy in the design of sustainable carbon materials. Creative approaches have been reported in the literature in recent years for the synthesis of green carbons that offer specific benefits compared with traditional carbons related to lower costs and greater design flexibility [1, 5–7].

Because of the rich chemical composition of the precursors used in the synthesis of green carbons (i.e., polysaccharides, lignocellulosic wastes, algae, tannins), their surface chemistry is very often much more complex than that of traditional carbon materials. This complexity defines the unique properties of these carbons and becomes one of the key aspects defining their performance in many areas of adsorption, catalysis and so on [3, 8]. These issues will be briefly addressed here and in other chapters of this book.

The surface chemistry of green carbons is, therefore, a topic of increasing interest among researchers who devote their activity to explore the new opportunities offered in different multidisciplinary areas. Although numerous reviews on the characteristics of the surface chemistry of conventional carbon materials can be found in the literature [3, 4], the purpose of this chapter is to provide a global view on the analysis of the surface chemistry of carbons along with the most commonly used techniques for its characterization.

1.2 Surface Functionalities

Carbon materials are mainly composed of sp^2-hybridized carbon atoms arranged in a two-dimensional structure and grouped in layers of fused aromatic rings (graphene sheets) with varied degree of planarity. These graphene layers are stacked by weak Van der Waals forces in either an ordered (i.e., graphite and graphitizable carbons) or turbostratic—disordered stacking—structure (i.e., activated carbons, carbon blacks) [9]. From structural and chemical viewpoints, a clear distinction can be made between basal-plane and edge carbon atoms in the graphene layers. The reactivity of carbon surfaces is defined by the basal-to-edges ratio; the edges have higher reactivity than the basal planes, and it is associated to chemisorption of heteroatoms giving rise to stable surface functionalities [10].

The chemical nature of the graphene edges and basal planes in sp^2-hybridized carbon atoms—a largely ignored and controversial issue—has been recently revisited by Radovic and co-workers [11, 12]. The main chemical features of the graphene sheets are (i) the surface functionalities either located at the edges (predominantly) or inserted in the basal-plane and (ii) the free edge sites. These free sites are associated to armchair (carbyne-like) and zigzag (carbene type) configuration carbon atoms (rather than H-terminated or free radicals as generally assumed for a long time) and account for the surface reactivity of carbons to chemisorb heteroatoms or in gasification reactions [12].

The concentration and relative distribution of the surface functionalities of carbons depend strongly on the material's origin, thermal history, any chemical post-modification, and the local environment of the surface groups. In the following sections, special emphasis is paid to the surface functionalities commonly present on green carbons upon the material's origin and processing. A brief review of common analytical techniques that allow an accurate characterization of carbons is also addressed.

1.2.1 Oxygen-Containing Functionalities

Oxygen-containing functionalities are by far the most common surface groups present on all types of carbon surfaces, mainly due to strong affinity of carbons to interact with oxygen when exposed to air (O_2 chemisorption occurs even at low temperature) [9, 11], although they can also be incorporated from the chemistry of the precursor. Most common structures corresponding to oxygen functional groups are shown in Fig. 1.1. For some of the schematized functions such as quinone and pyrone type, some other arrangements can also be proposed.

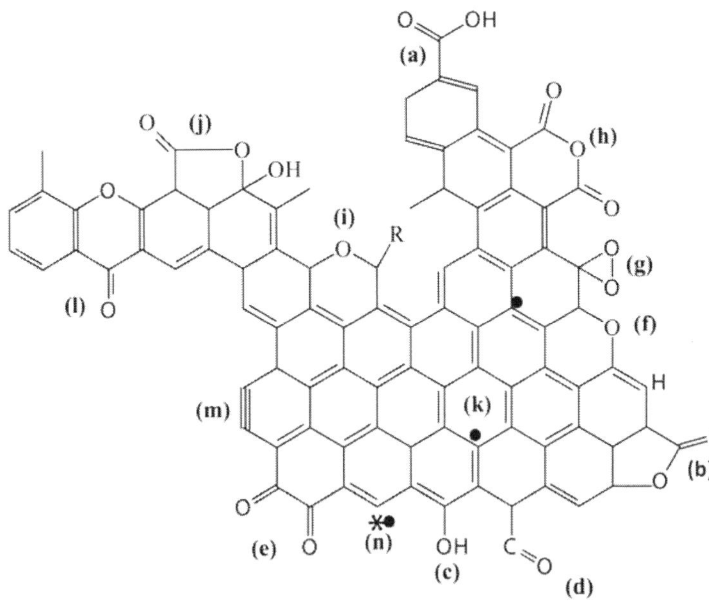

Figure 1.1 Schematic representation of the typical O-containing surface groups of carbon materials, including free edge sites and carbon–hydrogen complexes: (a) carboxyl groups, (b) lactone, (c) hydroxyl, (d) carbonyl, (e) quinone, (f) ether, (g) epoxyde, (h) carboxylic anhydride, (i) chromene, (j) lactol, (k) π electron density on carbon basal planes, (l) pyrone, (m) carbyne, and (n) carbene sites.

Oxygen functionalities are generally classified as acidic or basic based on the concept of acidity of surface groups introduced by Steenberg in 1944 [13]. Whereas the type of surface functionalities contributing to the acidic character of carbon materials is well documented (i.e., carboxyl groups, anhydrides, hydroxyls, lactones, and lactol groups), the understanding of carbons basicity is still rather controversial [14, 15]. Chromene, pyrone, and quinone-like structures, as well as non-heteroatomic Lewis base sites defined by regions of π electron density on the basal planes [16–18] have been postulated as main contributors to the carbons basicity, but their strength is still subject to debate [12].

The amount of naturally occurring oxygen in most carbons is rather weak (i.e., between 1 and 6 wt%) and mainly in the form of thermally stable surface oxides of basic nature. Acidic functionalities are more unstable and decompose at the high-temperature treatments [19] typically used during the synthesis. Notwithstanding the final oxygen content of carbons can be easily increased (ca. above 20 wt%) by post-modification treatments (oxidation *via* wet or gas phase) [20]. Unlike conventional carbons, the oxygen content in green carbons is much higher since the precursors usually present large oxygen quantities (Fig. 1.2). Moreover, the application of low-temperature synthesis conditions (i.e., hydrothermal carbonization) to transform these precursors into carbon-like materials also favors the development of a rich surface chemistry with a high oxygen content [7, 21].

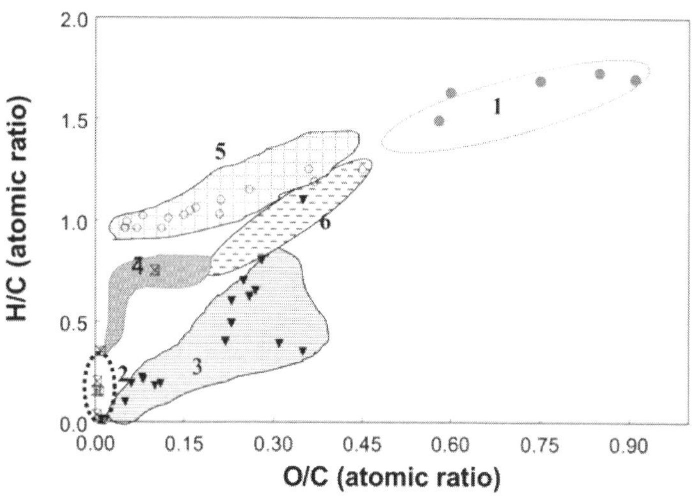

Figure 1.2 Van Krevelen diagram adapted from references [7, 21]. (1) raw biomass; (2) anthracites; (3) cellulose-derived chars from dry and hydrothermal pyrolysis; (4) bituminous coals; (5) digested chars; (6) lignites.

The type and distribution of acidic and basic oxygen groups present on a carbon surface depend on the synthetic route, the nature of the precursor, and the oxidation method applied if these are incorporated by oxidation procedures. Oxidative modifications are well documented in the literature and extensive work has been devoted to obtain acidic and basic carbons with different distributions of oxygen-containing functionalities upon the type and strength of the oxidizing agent [22–25]. It is well established that oxidation increases the acidity of carbons [24–26] and that both the nature of the groups and their distribution depend on the oxidation method. For instance, it has been reported that, as a general rule, gas phase oxidation increases mainly the concentration of hydroxyl and carbonyl surface groups, while wet oxidation mainly incorporates carboxylic acids [27, 28]. Moreover, oxidation using nitric acid incorporates higher concentrations of carboxylic groups and a more homogenous oxidation of carbon particles than ammonium persulfate [25, 29]. On the other hand, basic oxides are formed generally by decomposition of acidic groups at high temperatures, or after oxygen chemisorption upon re-exposure to air [30, 31].

Besides post-synthesis modifications, the preparation of green carbons using sustainable resources and low-temperature synthetic routes has generated interesting materials with singular oxygen–carbon complexes [1, 6, 7]. For instance, carbon produced from lignocellulosic wastes and other precursors containing ester, furanic, and/or phenolic functionalities as structural units [32–34] present high oxygen contents. During thermochemical transformation of these precursors, oxygen is inserted within the carbon matrix creating Lewis basic sites (mainly ether and chromene-like structures) and fair amounts of phenol-type and carbonyl groups. Carbons obtained from hydrothermal carbonization using polysaccharides and lignocellulosic precursors have also shown a unique structural and chemical composition (Fig. 1.3) [6, 7].

Although oxygen is uniformly distributed along these carbons, two distinct regions in a core–shell configuration are clearly differentiated from a chemical point of view (Fig. 1.3) as the nature of the oxygen-functional groups differ between the core and the shell of the particles [7, 35].

The core is hydrophobic and presents high oxygen content due to cross-linked furanic and ether or keto-aliphatic chains, rather than the aromatic structure of conventional and other biomass related carbons. In contrast, the hydrophilic shell contains a high density of oxygen surface groups of acidic nature (i.e., hydroxyl, carbonyl, or carboxylic). Such unusual chemical composition makes these materials suitable for their use in catalysis, drug delivery, enzyme immobilization, or selective and reactive adsorption [35, 36].

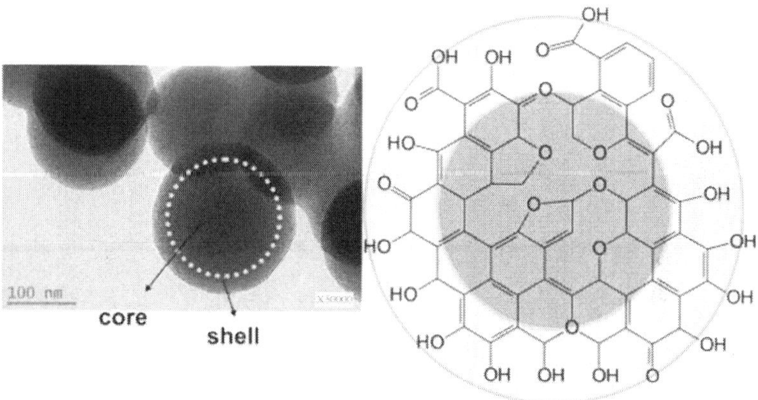

Figure 1.3 Schematic representation of the dual core–shell chemical composition of carbons obtained from hydrothermal carbonization of polysaccharides and lignocellulosic precursors and wastes.

1.2.2 Nitrogen Containing Functionalities

A long-coveted piece in the portfolio of carbon materials is the synthesis of nitrogen-doped carbons. Contrary to oxygen, carbon–nitrogen complexes are usually scarce in carbon surfaces as the amount of naturally occurring nitrogen in common precursors (coal or wood) is usually low. Nitrogen is essentially introduced into the carbon matrix in two ways, either by the carbonization of N-containing precursors or by post-modification methods; here, the carbon material is treated at a given temperature in the presence of N-containing species (i.e., ammonia, urea, melamine, cyanide, polyacrylonitrile, and so forth), which decompose leading to the formation of the desired N-surface groups [37–40]. Creative approaches have been recently reported for the preparation of N-doped carbons based on naturally (and sustainable) nitrogen containing precursors such as amino-carbohydrates or other N-enriched polymers (glucosamine or chitosan), aminoacids, proteins, N-ionic liquids, waste crab shells [35, 41–43].

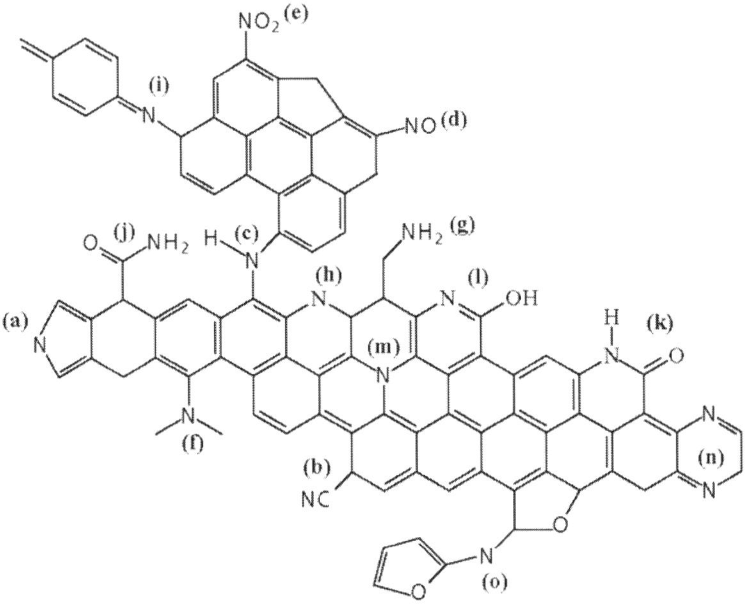

Figure 1.4 Schematic representation of nitrogen-containing groups: (a) pyrrole- like group, (b) nitrile, (c) secondary amine, (d) nitro group, (e) nitroso group, (f) tertiary amine, (g) primary amine, (h) pyridine-like group, (i) imine, (j) amide, (k) lactam, (l) pyridone, (m) quaternary amine, (n) pyrazine, and (o) nitrogen bridge in furanic groups.

Figure 1.4 shows a scheme of the different nitrogen-containing functionalities that may be formed on the carbon surfaces. Nitrogen atoms or N-containing functional groups are mostly bound to the graphene edges analogously to the surface oxides (i.e., in the form of amino or cyano groups). It is also possible to find nitrogen atoms inserted within the basal planes in substitution of C atoms in the form of quaternary nitrogen, or pyridine and pyrrole-type structures. The type and distribution of the N-groups largely depend on the temperature of the synthesis due to the different thermal stability of these groups (Fig. 1.5). Low-temperature treatments result in the formation of lactams, imides, and amines [44]. In contrast, high-temperature treatments favor the insertion of nitrogen atoms inside the graphene sheets in the form of quaternary nitrogen, pyridine, pyrrole-type and pyridine-N-oxide functionalities [45].

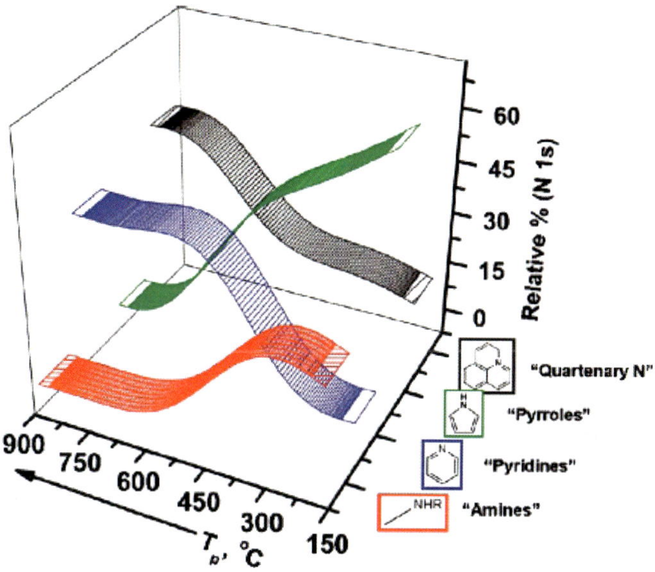

Figure 1.5 Evolution of N-containing groups on N-doped carbon aerogels obtained from glucose and ovalbumin with the carbonization temperature [43] (http://dx.doi.org/10.1039/c1gc15349h). Reproduced by permission of the Royal Society of Chemistry.

The overall acidic or basic character of N-doped carbons is also largely modified by the incorporation of nitrogen atoms in the carbon matrix. As a general rule, N-doping increases the basicity of carbons, although this depends on the type of incorporated groups and their distribution on the surface [46].

The nature of the functionalities also depends on the method applied to incorporate nitrogen groups. For instance, it has been reported that lactams and imides structures are the main functionalities formed by ammination, whereas amides are preferentially formed upon ammoxidation, though they may be transformed to pyrrole and pyridine by heat treatments [47]. On the contrary, amino groups are quite thermally unstable and thus their incorporation to carbon surfaces is usually carried out by low-temperature treatments or non-covalent immobilization *via* impregnation using liquid amines [48].

Studies on the structural incorporation of this heteroatom on different carbon matrices have revealed that it is a key parameter to control the performance in many applications such as the adsorptive properties of carbon materials [46], the activity in heterogeneous catalysis in basic reactions, dispersion of carbon-supported catalysts [49], or the electrochemical performance as electrodes [40, 43, 50]. For instance, the pseudocapacitive contribution of pyridinic and pyrrolic groups to the electrochemical performance of carbon electrodes, and the positive effect of quaternary and pyridinic-N-oxides as improved electron transfer groups [51, 52] has been reported.

1.2.3 Sulfur-Containing Functionalities

Sulfur is naturally present in carbon materials as elemental sulfur, inorganic species, and organo-sulfur compounds, with a content varying typically between 0–5 wt%. The incorporation of sulfur-containing surface groups is carried out by reaction of the carbon matrix with H_2S, CS_2, or SO_2 at various temperatures [53], or in the presence of elemental sulfur [54]. Typical sulfur contents of modified carbon surfaces are ca. 10 wt%. Most likely carbon–sulfur configurations present on carbon surfaces are shown in Fig. 1.6. These include sulfide and disulfides, thiophenols and thioquinones (formed by substitution reactions involving quinone and phenolic groups), and sulfoxide and sulfonic acid groups.

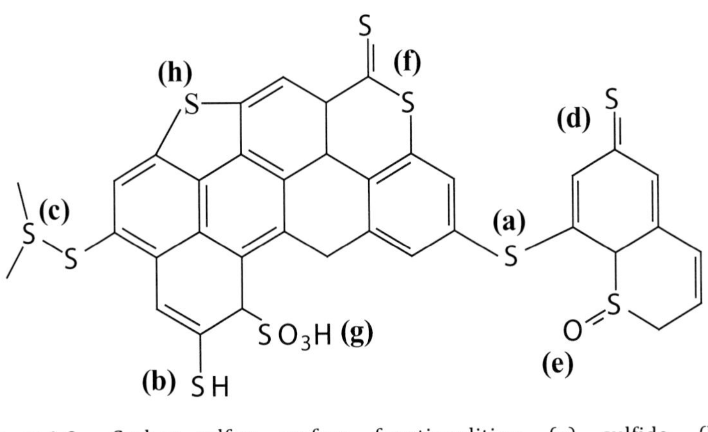

Figure 1.6 Carbon–sulfur surface functionalities: (a) sulfide, (b) thiophenol, (c) disulfide, (d) thioquinone, (e) sulfoxide, (f) thiolactone, (g) sulfonic, and (h) thiophene.

Carbon–sulfur complexes have been investigated for different applications such as water vapor adsorption on charcoals [53, 54], selective removal of metallic species (i.e., cadmium, mercury, lead) from liquid or gas phase [55], and removal of organo-sulfur pollutants in desulfurization applications [38, 56, 57], among most representatives. In this regard, the activity of the carbon–sulfur complexes relies on the selectivity of the adsorption process due to the ability of the target species (i.e., metal or organo-sulfur compound) to form strong bonds with the carbon–sulfur complexes [56–58].

1.2.4 Phosphorus-Containing Functionalities

Phosphorus is not a commonly encountered element in carbons, although it is present in carbons obtained using phosphoric acid activation [59–61]. In these reactions, the formation of phosphate and polyphosphate bridges provokes the expansion and cross-linking of the carbon matrix, driving to an accessible pore structure after the removal of the acid. The chemical state of phosphorus in P- containing carbons is a rather controversial issue [62, 63]. Most likely phosphorus-containing group son carbons proposed in the literature are sketched in Fig. 1.7.

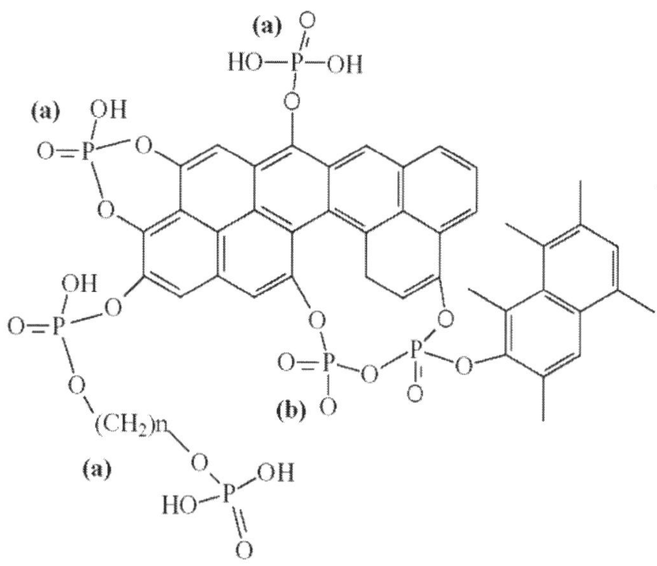

Figure 1.7 Schematic representation of the typical phosphorus functionalities: (a) phosphocarbonaceous esters and (b) pyrophosphate species.

Some experimental evidences using different analytical techniques (FTIR, XPS, NMR) have shown that the most abundant P species introduced in carbons by phosphoric activation are -C–P- or -C–O–P bonds [60] in phosphate and phosphonate-like structures. The presence of pentavalent phosphorus and elemental phosphorus is very rarely detected, unless high temperatures are applied [61].

Besides the generation of porosity using phosphoric acid, interesting photoelectrical properties have been reported for P-doped amorphous carbon films [64, 65], offering promising results for future prospects of clean and low-cost carbon solar cells. Recent studies have reported that P-containing groups might be important for the development of graphitic crystallites at 3273 K, which contrasts with so far reported role of P as inhibitor of carbon graphitization [66]. Moreover, phosphorus–nitrogen and phosphorus–carbon complexes have an important role on the electrochemical performance of carbons and the selectivity towards the adsorption of organo-sulfur compounds [51, 57, 67].

1.2.5 Halogens, Boron and Metallic Doping

Besides oxygen, nitrogen, sulfur, and phosphorus, some other heteroatoms can be successfully incorporated to carbons. Tackling the characteristics of all of them would merit a longer review, which cannot be attained in this chapter for space limitations. However, the next section addresses very briefly the main research aspects related to some other heteroatoms incorporated to carbon materials, which are attracting the attention of many researchers owing to their promising application to current technological and environmental problems.

1.2.5.1 Halogens

The treatment of carbon materials with halogen vapors or aqueous solutions of halogens results in the fixation of carbon–halogen complexes. Similar to other heteroatoms, the amount of halogen fixed depends on the carbon itself and the temperature of the treatment; the reactivity and stability of the carbon–halogen complex follows the order chlorine > bromine > iodine [4, 68]. Halogen–carbon groups are of great interest for further functionalization of carbon nanostructures (i.e., nanotubes) and in some applications.

For instance, some chlorine and iodine-impregnated carbons have been investigated as a cost-effective technology for the reduction of global emissions of mercury [69]. Also, mild fluorination of carbons seems to be a useful method to modify the surface polarity (hydrophobicity) of carbons as well as to produce fluorine–carbon composites as functional coating materials [70].

1.2.5.2 Boron

Highly boron-doped carbons (ca. 17 wt% boron) have been synthesized by standard CVD process using BCl_3. The substitutional doping of carbon atoms in sp^2 and sp^3 configuration with boron can modify the electronic and structural properties of the resulting carbon. Substituted boron atoms in the carbon lattice accelerate the graphitization and suppress the oxidation of carbon materials [71], which seems promising for their use as reinforcement materials in aerospace applications [72]. The positive effect of boron-doping on diamond and carbon electrodes [73] and in the field of hydrogen storage has also been reported [74], although further optimization of the local boron doping environment seems yet to be needed.

1.2.5.3 Hydrogen

Hydrogen is present in almost all forms of carbons either directly bonded to carbon atoms in aliphatic or aromatic environments or as a part of surface groups (ca. carboxylic acids, amines) and chemisorbed water. The hydrogen content can influence the degree of graphitization and their electrical resistance [75] in carbon blacks. Additionally, hydrogen treatment at high temperatures is used for the removal of surface functionalities under reductive atmosphere. Compared with thermal decomposition under inert atmosphere, the treatment under hydrogen provides carbons with a hydrophobic and stable surface, due to the stabilization of the highly reactive unsaturated carbon atoms generated at the edges of the graphene sheet [15]. As a result, obtained carbons do not undergo re-oxidation and formation of new oxygen groups when exposed to air.

1.2.5.4 Metallic doping

Immobilization of metals with catalytic activity (i.e., Pt, Co, Ni) on carbon supports is known to improve the stability, efficiency, and dispersion of the catalyst [2, 3]. The preparation of metal-doped carbons is usually performed by impregnation (wet and incipient-wetness) or ion-exchange methods, often followed by calcination and reduction treatments. The role of oxygen groups and basic sites on the dispersion of metals on carbon supports is widely documented in the literature, and as a general rule, oxidation leads to better dispersion and strong anchoring of the metallic species [76–78]. However, basic groups are more beneficial than acidic ones, and owing to the low thermal stability of acidic groups (i.e., carboxylic), sintering of the metallic particles might occur if high-temperature reduction treatments are applied [76–78]. Recently, an elegant way to obtain Ti-carbon catalysts has been reported using carbons prepared by hydrothermal carbonization [79], where the incorporation of the metal is controlled by the hydroxyl groups located at the carbon shell. The obtained core–shell carbon titania composites posses controlled morphology and topology of the active phase, and enhanced photocatalytic activity for degradation of pollutants [80, 81].

1.3 Characterization Techniques

The characterization of the surface chemistry of carbon materials is a complex issue, and no single analytical chemistry discipline is able (not recommended) to provide an accurate and successful characterization of carbon surfaces. A global and coherent picture of all surface chemical properties of carbons can only be obtained when several techniques are used, and complementary information of different physicochemical aspects is retrieved.

Some of these techniques provide qualitative information on the type and distribution of groups on the carbon, such as infrared and X-ray photoelectron spectroscopy (XPS), nuclear magnetic resonance and potentiometric techniques; some others, such as Boehm titration and thermal analysis, provide quantitative information about the carbon surface. Owing to space limitations, only a brief review of selected analytical techniques is presented in this section. All of them are commonly employed research tools to determine the composition, distribution, nature, and local environment of the carbon surface. Detailed information on a larger number of techniques can be found in the literature [3, 4].

1.3.1 Thermal Analysis

Temperature-programmed desorption consists of heating of the carbon in a flowing carrier gas at a controlled heating rate to induce the thermal decomposition of the surface functionalities chemically bound to carbons. The evolved gases can be analyzed by various techniques (mass spectrometry, IR, gas chromatography). CO, CO_2 and H_2O are the dominant desorbed gases, although some other oxides (NO, N_2, HCN, NH_3, SO_2) can also be detected in N- and S-doped carbons. The correlation between the gaseous species detected upon thermal decomposition (the number and temperature of the desorption peaks) and the nature of the carbon surface functionalities has been widely studied [19, 28, 82–84]. TPD profiles due to the decomposition of carbon-heteroatoms functional groups are usually characterized by wide humps. Occasionally some sharp peaks associated to the degradation of the inorganic content of the carbons might appear, and therefore the assignment to surface functionalities must be cautiously done [85].

Deconvolution of the TPD spectra is a useful tool to identify different groups [19, 28, 84]. It is assumed that complexes decomposing as CO_2 can be split into carboxylic acids, anhydrides, and lactones. Similarly, the CO-evolving groups account for anhydrides, phenols and ethers/carbonyls structures (Fig. 1.8). In the case of N-doped carbons, it is difficult to assign the pristine functional group based only on the desorption profiles, although some pyridine, imine-type, linear and cyclic amide, and nitrile groups have been suggested [84].

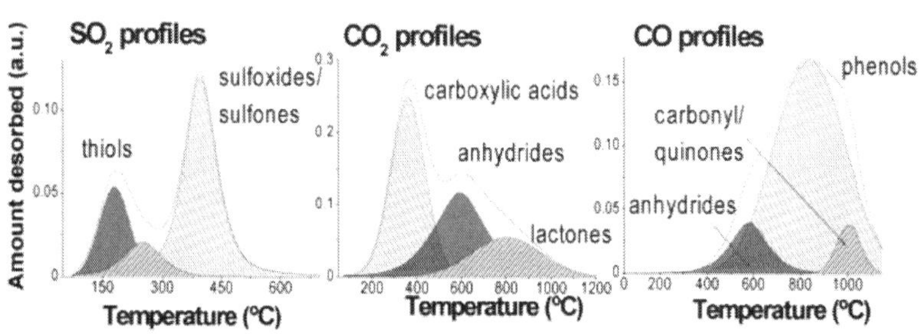

Figure 1.8 Deconvolution of TPD profiles of SO_2-, CO_2-, and CO-evolving groups on an activated carbon obtained from a plastic waste and modified with ammonium persulfate.

Sulfur-containing functionalities mainly decompose as H_2S and SO_2; thiol groups have been reported to decompose at low temperature (between 573 and 613 K) as H_2S and SO_2—upon recombination with surface oxygen during the thermal decomposition. The peaks appearing between 633 and 700 K have been attributed to sulfonic acids, sulfoxides and/or sulfone and C–S–C sulfides [86]. In general, the results obtained by TPD agree well with the information obtained from other techniques and contribute to a more accurate picture of the surface chemistry of carbons.

1.3.2 X-Ray Photoelectron Spectroscopy

X-ray photoelectron spectroscopy is a non-destructive quantitative technique that provides information on the chemical state and local environment of the atoms. It is based on the use of X-rays (ca., 1–2 eV) to excite the electrons from the inner shell of the atoms, which are emitted with a characteristic kinetic energy, dependent on the incident photon energy and the binding energy of the atoms. The intensity of the peaks correlates to the concentration of the analyzed element, rendering quantitative analysis of the surface of the solid. Additionally, it allows the evaluation of the surface distribution of the functionalities based on the fact that XPS analysis accounts for the heteroatoms on the external surface whereas other techniques account for the overall distribution on the bulk material (i.e., elemental or thermal analyses) [31]. This indicates whether the functionalities are homogeneously distributed on the carbon or preferentially located in the external surface.

XPS is used to characterize surface functionalities on carbons based on the assignments of the binding energies of photoelectrons (i.e., C1s, N1s, O1s, P2p, S2p, and so forth) to surface groups [47, 86]. The assignments corresponding to the main groups present in carbons are compiled in Table 1.1. The assignment is largely affected by the local environment of the heteroatoms, the state of bonding to neighboring atoms, etc, for which identification is not straightforward. Despite this, XPS analysis can be very convenient to identify any likely changes on the surface of carbon materials during pyrolysis/carbonization treatments [41, 43], or those arising after long-term uses [40].

Table 1.1 Binding energies (eV) assigned to surface functionalities on carbon materials [45, 86, 87]

Peak	Component	Position (eV)	Function
C_{1s}	Csp^2	284.2–284.9	Aromatic carbon (graphite)
	C*–COOR	284.6–285.5	α Carbon of carboxyl groups
	C*–OR	285.4–286.1	Ether and hydroxyl (phenol) group
	C*=O	287.2–287.9	Ketone groups
	C*OOR	288.4–289.5	Carboxyl groups
	Sat	289.9–291.6	Satellite of C*=O component
	Π–Π*	290–291.7	Π–Π* transition of aromatic carbons
	C*–S	285.1–285.8	Carbon linked to sulfur
	C*–O–P	285.5–286.3	Carbon in phosphorus group
	C–N	286.3–287.5	Carbon in nitrogen group
O_{1s}	C=O*	529.5–532.5	Oxygen in ketone groups
	C=O* carboxyl	530.1–533.1	C=O in carboxyl groups
	C–O*–C	533.2–533.8	Ether groups
	C–O*R	531.7–534.7	Bridging oxygen
	H_2O	533.7–536.7	Water physisorbed, chemisorbed
	S(O*)	529.5–530.6	Oxygen linked to sulfur
	O*=P	530.9–532.5	Oxygen double linked to phosphorus
	O*–P	530.9–532.6	Oxygen linked to phosphorus, C–O–P, O–P
P_{2p}	PO_4^{3-}, $P_2O_7^{4-}$	132.5–133.5	Phosphate and pyrophosphate
	PO_3^-	134.3–134.6	Metaphosphate
	P_2O_5	135.6–136	P_2O_5
S_{2p}	–SH, CS_2	163.5–163.7	Thiol or carbon bisulfide
	C–S–C	164.5–164.7	Sulfides or thioesters
	$R_2S=O$, SO_3^{2-}	167.0–167.2	Sulfoxides or sulfites
	$R–SO_2–R$	168.0–168.5	Sulfones
	SO_4^{2-}	168.4–168.8	Sulfates
	$R–SO_3H$	169.5–169.7	Sulfonic acids
N_{1s}	C–N–C	399.9–400.7	Pyrrole, pyridone
	$C–N^+H–C$	401.1–401.7	Quaternary nitrogen
	N–O–C, $–NO_2$	405.5–406.1	Oxidized nitrogen, $–NO_2$
	C–N–C	398.6–398.9	Pyridine
	$–NH_2$, –NH, C=N	399.7–400.8	Imine, amide, amine, lactame
	C–N–Ox, NH_3	402.6–402.9	Pyridine-N-oxide or ammonia
	CN	399.4–400.1	Nitrile
	$C–N^+$	402.2–403.1	Protonated pyridine, pyrrole, lactams, and amide

1.3.3 Nuclear Magnetic Resonance

This spectroscopic technique relies on the magnetic properties of atomic nuclei exposed to electromagnetic radiation in the frequency region 4–900 MHz. Not frequently applied to study the surface of carbons, NMR has recently become more popular, particularly for the study of the transformations occurring in carbon matrices during thermal decomposition of various precursors [35, 61, 88], or the evaluation of changes in surface chemistry of carbons [89].

Advanced solid state NMR techniques based on deconvolution of the spectral parameters of carbon and other heteroatoms (^{13}C, ^{1}H, ^{15}N, ^{31}P, ^{11}B, and ^{19}F), can give the ratio of aromatic to aliphatic carbon in the structure, and the heteroatom arrangements present on the carbon surface. For instance, ^{13}C, ^{1}H, and ^{15}N NMR spectra have confirmed the presence of polyfuranic structures in the internal structure (core) of carbons prepared from hydrothermal carbonization, rather than the aromatic structure of conventional carbons [35]. For N-doped carbons, the presence of pyrazine motifs connected to furans, and the quasi-absence of NH_2 groups has also been evidenced [43]. In the case of P-doped carbons, the large chemical shift for ^{31}P allowed to distinguish various phosphorus-containing compounds [57, 60, 61]. Another NMR approach for evaluating the chemical nature of carbon surfaces is based on adsorption of ^{129}Xe [90, 91]. The interpretation of the results relies on the analysis of chemical shift in the absorption frequency of xenon correlated to the density of surface functionalities of the carbon. This dependence has been reported for oxygen and chlorine containing groups.

1.3.4 Electrochemical Techniques

As other physicochemical features of carbon materials, the electrochemical behavior of carbons—the double-layer capacitance, wettability, charge and mass transport, potentiometric response of carbon to ionic species in solution is strongly influenced by the attachment of functionalities on the surface. Those readers interested in a deep discussion on the application of carbons in electrochemical systems and the study of the electrochemical phenomena at active carbon-electrolyte solution interfaces are referred to the comprehensive reviews by Kinoshita [92] and Biniak [93].

A number of experimental techniques can be used for determining the electrochemical performance of carbon materials: voltammetry, polarography, impedance spectroscopy. Typically three electrode configuration cells are used, where the carbon material is a component of the working electrode, and its performance is analyzed in a fixed range of potential versus a reference electrode [94, 95]. Various electrolytes covering aqueous (i.e., H_2SO_4, Na_2SO_4, KOH, NaCl) and non-aqueous (organic salts dissolved in aprotic solvents) media can be used. Wide anodic and/or cathodic peaks or humps on cyclic voltammograms of carbon electrodes reveal the presence of redox couples of electroactive species, as shown in Fig. 1.9. The waves can be rather unevenly shaped and spread over a wide potential range when various redox processes overlap [40, 92], which creates some difficulties for the identification of the functionalities present on a carbon material.

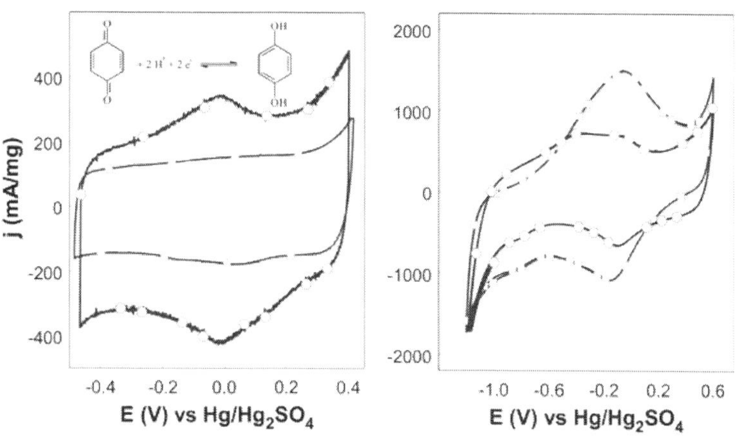

Figure 1.9 Current-potential profiles for various activated carbons. Left: pristine (squares) and after wet oxidation (circles). Right: raw oxygen-containing carbon (circles) and after nitrogen incorporation (squares). Scan rate 2 mV/s; 1M H_2SO_4 electrolyte.

Functional groups with heteroatoms bearing an unshared pair of electrons (i.e., oxygen, nitrogen, sulfur, halogens) may be involved in redox reactions [17, 92, 96]. Highly reversible electrochemical redox systems involving oxygen, nitrogen and phosphorus have been reported in the literature: the reduction potential of the quinone/hydroquinone pair in different environments can appear at values close to −0.15 V versus Hg/Hg_2SO_4 in carbon materials [92, 93]; pyrone-type structures can also appear in the same electrochemical potential range [17,18]. For nitrogen functionalities, a number of mechanisms involving amides, amines and pyridine-like structures have been proposed [50–52, 97, 98].

Electrochemical techniques have also been used to identify modifications of surface functionalities upon long-term use of the carbon material [99] or to functionalize the surface. In this regard, it has been reported that both anodic and cathodic treatments can lead to an increase in the surface oxygen groups. Moreover, a suitable choice of the electrochemical variables might allow a selective modification of the carbon surface [100].

The electrochemical characterization of surface groups becomes quite complex when different electro-active heteroatoms are present on the carbon surface. Moreover it is a qualitative technique, although a few attempts of quantification have been made by means of voltammetry and polarography [101]. However, the information about the type of surface groups is in good agreement with characterization obtained using other techniques [99, 100].

1.3.5 Infrared Spectroscopy

With the development of Fourier analysis, infrared spectroscopy (FTIR) has become a popular and frequently used characterization technique to analyze the surface chemistry of materials, including carbons [4, 93, 102]. The main challenge in FTIR analysis of carbon materials is that most of them are too opaque for direct transmission analysis in the mid-infrared spectral region. Partial dilution in an appropriate transparent additive (KBr) is not always satisfactory, for which alternative sampling techniques such as diffuse reflectance, photo-acoustic spectroscopy, or attenuated reflectance are extensively used to overcome these problems [102, 103]. IR spectroscopy does not provide quantitative information, and the assignment of the absorption bands to the surface groups on carbons is made by comparison with the IR spectra for organic compounds with similar functionalities. Table 1.2 shows the assignments to oxygen, nitrogen, sulfur and phosphorus-containing groups according to literature [7, 43, 86, 104, 105]. Since the frequency of the absorption bands can be affected by the local environment of the surface groups through resonance and conjugation effects with

Table 1.2 Infrared absorption bands on carbon surfaces and their corresponding assignments to surface functionalities [87, 102, 103, 105]

Group	Assignment regions (cm^{-1})	Group	Assignment regions (cm^{-1})
C–O stretch of ethers	1000–1300	N–O, amine oxide	940–980, 1200–1300
C–O–C, ether bridge	1230–1250	=N–OH	930–960, 1650–1680, 3550–3600
-OH	1049–1276; 3200–3640;	SH, thiols	2550–2600
-O–H bend/stretch	1160–1200; 2500–3620	S–OR, esters	700–900
Aromatic C=C stretching	1585–1600	S–S, disulfide	500–540
C=O, quinones	1550–1680	C=S thiocarbonyl	1050–1200
Carboxylic acids	1120–1200; 1665–1760; 2500–3300	S=O sulfoxide,	1030–1070
Carbonates, carboxyl-carbonates	1000–1500; 1590–1600	S=O sulfone,	1320–1350, 1120–1160
Ketenes (C=C=O)	2600–3000	S=O sulfonic acid	1250–1345
Lactones	1160–1370; 1675–1790	SO_4^{2-}	1350–1450
Anhydrides	980–1300; 1740–1880; 2080–2200	P–OR esters	900–1050
C=C, C=N, NH_3	1620–1650	P=O, phosphate, phosphoramide	1100–1300
CONH, cyclic amides	1546	C–O–P stretching in phosphate esters	1064, 1065
-CN, nitrile	1254	P–OH stretching	970, 918
sp^2 carbon, C–N, C=N	1300–1350	P–OH stretching in phosphate esters	831
Raman G, C=N	1550–1570	Amines	1200
C–N, C=N, C=C	1500–1510	Thiols	2500–2600
C–N (aliphatic), N–H	1020–1150	(O)=PO–H, phosphonic acid	2550–2700
N=O, nitro, nitroso	1550–1530, 1320–1380	P–H phosphine	2280–2440, 950–1250

neighboring groups, a univocal assignment is not always achieved. For instance, quinone groups are generally assigned either to the band at 1550–1680, or 1660–1670 cm^{-1}; the band at 1660 cm^{-1} has also been assigned either to aromatic C=C stretching vibration of the basal planes and to oxygen surface compounds [86, 102]. Moreover, FTIR is a convenient method for the identification of NH_2 groups, which are difficult to detect, for instance, using XPS since their N1s binding energy overlaps that of pyridine and cyano-like groups [104].

1.3.6 Other Techniques

1.3.6.1 Elemental analysis

Quantification of the heteroatoms is usually expressed in terms of mass fraction of the elements. Carbon, hydrogen, nitrogen and sulfur contents are determined by combustion of the carbon in excess of oxygen at high temperatures (i.e., 1473 K). In a complete combustion, all carbon is converted to CO_2, H_2O, SO_2, and N_2 (previous reduction by copper), which are collected and analyzed; the solid residue renders the ash content. The constituents of the inorganic matter are analyzed by atomic spectroscopy techniques, upon digestion of the ashes in strong acids.

1.3.6.2 Boehm titration

This classical analytical chemistry approach is one of the most popular techniques used to characterize the carbon surface chemistry [22, 106]. The method is based on the selective neutralization of the surface groups using bases of increasing strength: $NaHCO_3$ ($pK_a = 6.37$), Na_2CO_3 ($pK_a = 10.25$), NaOH ($pK_a = 15.74$), and $NaOC_2H_5$ ($pK_a = 20.58$). It is assumed that sodium bicarbonate neutralizes carboxylic acids, sodium carbonate—carboxylic acids and lactones, sodium hydroxide—carboxylic acids, lactones and phenols, whereas sodium ethoxide reacts with all oxygen species. The main cornerstone of Boehm titration is the lack of selectivity, which is imposed by the values of pK_a of the surface species. Boehm method was initially developed for oxygen-containing groups; thus it fails to characterize carbons where besides oxygen other heteroatoms (i.e., nitrogen, sulfur, phosphorus) are present—very often the case of N-doped carbons [107]. Nevertheless, for carbon–oxygen complexes Boehm titration provides a reliable analysis and is still considered a powerful method of first screening for carbon surface chemistry and determining the general trend in surface acidity. By analogy with Boehm titration, methods to identify the different basic sites with acids of increasing strength have been proposed [13]. However, from a practical point of view, the total content of basic groups is generally determined by titration with HCl [34].

1.3.6.3 Potentiometric titration

This approach considers that carbon surface is composed of acidic sites characterized by different acidity constants, K_a, and that the population of sites can be described by a continuous pK_a distribution, f(pK_a). Experimentally, an aqueous suspension of the carbon in a neutral electrolyte is titrated with NaOH in the pH range 3–10 [108, 109]. The obtained titration curve can be transformed into a proton-binding isotherm representing the total amount of protonated sites, which is related to the pK_a distribution by an integral equation. The solution to the integral equation using an application of the numerical procedure SAIEUS (Solution of Adsorption Integral Equation Using Splines) [110, 111] allows one to obtain distributions of acidity constants where peaks are associated to the various functional groups of the carbon. This provides a precise estimation of surface groups having a certain pK_a value. The exact classification of species is not possible, as many heteroatom configurations can result in similar pK_a. This method also presents some limitations related to the slow establishment of the ion exchange equilibria, the sample's particle size, and the rate of titration [112]. However, when it is carried out under the optimized conditions, it is a *via*ble and independent method to describe in detail the acid–base chemistry of carbon surfaces in aqueous electrolyte suspensions [112].

1.3.6.4 pH, point of zero charge and isoelectric point

The specific identification of surface groups is sometimes less important that the determination of the acidic/nature of the carbon surface, as well as its surface charge. This can be readily measured by simple techniques as the pH, point of zero charge (PZC) and/or isoelectric point (IEP). Brönsted acidic groups of the carbon surface become negatively charged in solution as they donate their protons to water molecules. Lewis bases adsorb protons from solution, becoming positively charged. Thus, carbon surface functionalities are responsible for the amphoteric nature of carbon, the pH in aqueous solution, and the surface charge. The pH of aqueous slurry of carbons provides an average indication of the acidic/basic nature; it can be measured by the standard test proposed by the American Society for Testing Materials [113].

The electrical state of the carbon surface in solution can also be characterized by the PZC and IEP. The IEP is defined as the pH where the charge at the surface of the hydrodynamic shear of the Stern Layer is zero; it is generally measured by applying an electric field to move the charged carbon particles in a colloidal solution by microelectrophoresis or electroosmosis methods [15]. The PZC is defined as the pH where the net surface charge, resulting from the adsorption of the potential determining ions (i.e., H^+ and OH^-), is zero, and it can be determined by mass titration methods [114, 115]. The PZC is considered the pH value below which the surface of the carbon particles in solution is, on average, positively charged (pH > PZC) and vice versa. Whereas electrophoretic mobility measurements are only representative of the external surface charges of carbon particles, the PZC accounts for net total (external and internal) surface charge of the particles. Consequently, the PZC-IEP difference can be interpreted as a measure of the carbon surface charge distribution, with positive values corresponding to more positively charged internal than external particle surfaces (i.e., less acidic external than internal surfaces), and values close to zero corresponding to a more homogeneous distribution of the surface charges [15].

References

1. Hu, B., Wang, K., Wu, L. H., Yu, S. H., Antonietti, M., and Titirici, M. M. (2002). Engineering carbon materials from the hydrothermal carbonization process of biomass. *Adv. Mater.*, **2**, pp. 813–828.
2. Inagaki, M., Park, K. C., and Endo, M. (2010). Carbonization under pressure. *New Carbon Mater.*, **25**, pp. 409–420.
3. Serp, Ph., and Figueiredo, J. L. eds. (2009). *Carbon Materials for Catalysis* (Wiley & Sons, Inc. USA).
4. Bandosz, T. J. ed. (2006). *Activated Carbon Surfaces for Environmental Remediation* (Elsevier, USA).

5. Szczurek, A., Amaral-Labat, G., Fierro, V., Pizzi, A., Masson, E., and Celzard, A. (2011). The use of tannin to prepare carbon gels. Part I: Carbon aerogels. *Carbon*, **49**, pp. 2773–2784.

6. Lee, J., Kim, J., and Hyeon, T. (2006). Recent progress in the synthesis of porous carbon materials, *Adv. Mater.*, **18**, pp. 2073–2094.

7. Sevilla, M., and Fuertes, A. B. (2009). Chemical and structural properties of carbonaceous products obtained by hydrothermal carbonization of saccharides. *Chem. Eur. J.*, **15**, pp. 4195–4203.

8. Frackowiak, E., and Béguin, F. (2001). Carbon materials for the electrochemical storage of energy in capacitors. *Carbon*, **39**, pp. 937–950.

9. Henning, G. R. (1966). *Chemistry and Physics of Carbon* 2, ed. Walker, P. L., Chapter 1, "Electron microscopy of reactivity changes near lattice defects in graphite" (Marcel Dekker, New York), pp. 1–49.

10. Radovic, L. R., Walker, P. L., and Jenkins, R. G. (1983). Importance of carbon active sites in the gasification of coal chars. *Fuel*, **62**, pp. 849–856.

11. Radovic, L. R., and Bockrath, B. (2005). On the chemical nature of graphene edges: origin of stability and potential for magnetism in carbon materials. *J. Am. Chem. Soc.*, **127**, pp. 5917–5927.

12. Radovic, L. R., Silva-Villalobos, A. F., Silva-Tapia, A. B., and Vallejos-Burgos, F. (2011). On the mechanism of nascent site deactivation in graphene. *Carbon*, **49**, pp. 3471–3487.

13. Steenberg, B. (1944). *Adsorption and Exchange of Ions on Activated Charcoal* (Almquist & Wiksells, Uppsala) p. 59.

14. Leon-Leon, C. A., Solar, J. M., Calemma, V., and Radovic, L. R. (1992). Evidence for the protonation of basal plane sites on carbon. *Carbon*, **30**, pp. 797–811.

15. Menendez, J. A., Illan-Gomez, M. J., Leon-Leon C. A., and Radovic, L. R. (1995). On the difference between isoelectric point and point of zero charge of carbons. *Carbon*, **33**, pp. 1655–1657.

16. Fuente, E., Menendez, J. A., Suarez, D., and Montes-Moran, M. A. (2003). Basic surface oxides on carbon materials: a global view. *Langmuir*, **19**, pp. 3505–3511.

17. Montes-Moran, M. A., Suarez, D., Menendez, J. A., and Fuente, E. (2004). On the nature of basic surface oxides on carbon materials: a global view. *Carbon*, **42**, pp. 1219–1225.

18. Montes-Moran, M. A., Menendez J. A., Fuente, E., and Suarez, D. (1998). Contribution of the basal planes to carbon basicity: an ab initio study of the H_3O^+–π interaction in cluster models. *J. Phys. Chem. B*, **102**, pp. 5595–5601.

19. Figueiredo, J. L., Pereira, M. F. R., Freitas, M. M. A., and Orfao, J. J. M. (1999). Modification of the surface chemistry of activated carbons. *Carbon*, **37**, pp 1379–1389.

20. Ania, C. O., Parra, J. B., and Pis, J. J. (2002). Influence of oxygen containing functional groups on active carbon adsorption of selected organic compounds. *Fuel Process Technol.*, **79**, pp. 265–271.

21. Mumme, J., Eckervogt, E., Pielert. J., Diakité, M., Rupp, F., and Cern, J. (2011). Hydrothermal carbonization of anaerobically digested maize silage. *Biores. Technol.*, **102**, pp. 9255–9260.

22. Boehm, H. P. (1999). Some aspects of the surface chemistry of carbon blacks and other carbons. *Carbon*, **32**, pp. 759–769.

23. Salame, I. I., and Bandosz, T. J. (2001). Surface chemistry of activated carbons: combining temperature-programmed desorption, Boehm, and potentiometric titrations. *J. Colloid Interf. Sci.*, **240**, pp. 252–258.

24. Moreno-Castilla, C., Lopez-Ramon, M. V., and Carrasco-Marin, F. (2000). Changes in surface chemistry of activated carbons by wet oxidation. *Carbon*, **38**, pp. 1995–2001.

25. Domingo-García, M., López Garzón, F. J., and Pérez-Mendoza, M. (2000), Effect of some oxidation treatments on the textural characteristics and surface chemical nature of an activated carbon. *J. Colloid Interf. Sci.*, **222**, pp. 233–240.

26. Ania, C. O., Cabal, B., Parra, J. B., Arenillas, A., Arias, B., and Pis, J. J. (2008). Naphthalene adsorption on activated carbons using solvents of different polarity. *Adsorption*, **14**, pp. 343–355.

27. Domingo-García, M., López-Garzón, F. J., and Pérez-Mendoza., M. (2002). On the characterization of chemical surface groups of carbon materials. *J. Colloid Interf. Sci.*, **222**, pp. 116–122.

28. Haydar, S., Moreno-Castilla, C., Ferro-Garcia, M. J., Carrasco-Marin, F., Rivera-Utrilla, J., Perrard, A., and Joly, J. P. (2000). Regularities in the temperature-programmed desorption spectra of CO_2 and CO from activated carbons. *Carbon*, **38**, pp. 1297–1308.

29. Otake, Y., and Jenkins, R. G. (1993). Characterization of oxygen-containing surface complexes created on a microporous carbon by air and nitric acid treatment. *Carbon*, **31**, pp. 109–121.

30. Menendez, J. A., Radovic, L. R., Xia, B., and Phillips, J. (1996). Low-temperature generation of basic carbon surfaces by hydrogen spillover. *J. Phys. Chem.*, **100**, pp. 17243–17248.

31. Menendez, J. A., Phillips, J., Xia, B., and Radovic, L. R. (1996). On the modification and characterization of chemical surface properties of activated carbon: in the search of carbons with stable basic properties. *Langmuir*, **12**, pp. 4404–4410.

32. Mestre, A. S., Pires, J., Nogueira, J. M. F., Parra, J. B., Carvalho, A. P., and Ania, C. O. (2009). Waste-derived activated carbons for removal of ibuprofen from solution: role of surface chemistry and pore structure. *Bioresour. Technol.*, **100**, pp. 1720–1726.

33. Carrott, P. J. M., Ribeiro-Carrott, M. M. L., and Lima, R. P. (1999). Preparation of activated carbon "membranes" by physical and chemical activation of cork. *Carbon*, **37**, pp. 515–517.

34. Budinova, T., Savova, D., Tsyntsarski, B., Ania, C. O., Cabal, B., Parra, J. B., and Petrov, N. (2009). Biomass waste-derived activated carbon for the removal of arsenic and manganese ions from aqueous solutions. *Appl. Surf. Sci.*, **255**, pp. 4650–4657.

35. Baccile, N., Laurent, G., Coelho, C., Babonneau, F., Zhao, L., and Titirici, M. M. (2011). Structural insights on nitrogen-containing hydrothermal carbon using solid-state magic angle spinning ^{13}C and ^{15}N nuclear magnetic resonance. *J. Phys. Chem. C*, **115**, pp. 8976–8982.

36. Cabrita, I., Ruiz, B., Mestre, A. S., Carvalho, A. P., and Ania, C. O. (2010). Acetominophen removal on low cost activated carbons produced from residues. *Chem. Eng. J.*, **163**, pp. 249–255.

37. Arenillas, A., Rubiera, F., Parra, J. B., Ania, C. O., and Pis, J. J. (2005). Surface modification of low cost carbons for their application in the environmental protection. *Appl. Surf. Sci.*, **252**, pp. 619–624.

38. Seredych, M., Portet, C., Gogotsi, Y., and Bandosz, T. J. (2009). Nitrogen modified carbide-derived carbons as adsorbents of hydrogen sulfide. *J. Colloid Interf. Sci.*, **330**, pp. 60–66.

39. Raymundo-Piñero, E., Cazorla-Amorós, D., Linares-Solano, A., Find, J., Wild, U., and Schlögl R. (2002). Structural characterization of N-containing activated carbon fibers prepared from a low softening point petroleum pitch and a melamine resin. *Carbon*, **40**, pp. 597–608.

40. Ania, C. O., Khomenko, V., Raymundo-Piñero, E., Parra, J. B., and Beguin, F. (2007). Improved electrochemical performance in aqueous medium of doped-carbons obtained using a zeolite as template. *Adv. Funct. Mater.*, **17**, pp. 1828–1836.

41. Baccile, N., Antonietti, M., and Titirici, M. M. (2010). One-step hydrothermal synthesis of nitrogen-doped nanocarbons: albumine directing the carbonization of glucose. *Chem. Sus. Chem.*, **3**, pp. 246–253.

42. Gutierrez, M. C., Carriazo, D., Ania, C. O., Parra, J. B., Ferrer, M. L., and del Monte, F. (2011). Deep eutectic solvents as both precursors and structure directing agents in the synthesis of nitrogen doped hierarchical carbons highly suitable for CO_2 capture. *Energy Environ. Sci.*, **4**, pp. 3535–3544.

43. White, R. J., Yoshizawa, N., Antonietti, M., and Titirici, M. M. (2001). A sustainable synthesis of nitrogen-doped carbon aerogels. *Green Chem.*, **13**, pp. 2428–2434.

44. Elsayed, Y., and Bandosz, T. J. (2002). Acetaldehyde adsorption on nitrogen-containing activated carbons. *Langmuir*, **18**, pp. 3213–3218.

45. Biniak, S., Szymanski, G., Siedlewski, J., and Swiatkowski, A. (1997). The characterization of activated carbons with oxygen and nitrogen surface groups. *Carbon*, **35**, pp. 1799–1810.

46. Lahaye, J., Nanse, G., Bagreev, A., and Strelko, V. (1999). Porous structure and surface chemistry of nitrogen containing carbons from polymers. *Carbon*, **37**, pp. 585–590.

47. Kapteijn, F., Moulijn, J. A., Matzner, S., and Boehm, H. P. (1999) The development of nitrogen functionality in model chars during gasification in CO_2 and O_2. *Carbon*, **37**, 1143–1150.

48. Plaza, M. G., Pevida, C., Arenillas, A., Rubiera, F., and Pis, J. J. (2007). Post-combustion CO_2 capture with a commercial activated carbon: Comparison of different regeneration strategies. *Fuel*, **86**, pp. 2204–2212.

49. Zhao, L., Fan, L. Z., Antonietti, M., and Titirici M. M. (2010). Nitrogen-containing hydrothermal carbons with superior performance in supercapacitors. *Adv. Mater.*, **22**, pp. 5202–5206.

50. Hulicova-Jurkahova, D., Kodama, M., Shiraishi, S., Hatori, H., Zhu, Z. H., and Lu, G. Q. (2009). Nitrogen enriched nonporous electrodes with extraordinary supercapacitance. *Adv. Funct. Mater.*, **19**, pp. 1800–1809.

51. Hulicova, D., Yamashita, J., Soneda, Y., Hatori, H., and Kodama, M. (2005). Supercapacitors prepared from melamine-based carbon. *Chem. Mater.*, **17**, pp. 1241–1247.

52. Hulicova-Jurcakova, D., Seredych, M., Lu, G. Q., and Bandosz, T. J. (2009). Combined effect of nitrogen- and oxygen-containing functional groups of microporous activated carbon on its electrochemical performance in supercapacitors. *Adv. Funct. Mater.*, **19**, pp. 438–447.

53. Puri, B. R. (1970). *Chemistry and Physics of Carbon*, **6**. ed. Walker, P. L., "Surface complexes on carbons" (Marcel Dekker, New York) pp. 191–282.

54. Blayden, H. E., and Patrick, J. W. (1967). Solid complexes of carbon and sulfur-1. Sulphurized polymer carbons. *Carbon*, **5**, pp. 533–544.

55. Macias-Garcia, A., Valenzuela-Calahorro, C., Espinosa-Mansilla, A., Bernalte-Garcia, A., and Gomez-Serrano, V. (2004). Adsorption of Pb^{2+} in aqueous solution by SO_2^- treated activated carbon. *Carbon*, **42**, pp. 1755–1764.

56. Seredych, M., and Bandosz, T. J. (2011). Investigation of the enhancing effects of sulfur and/or oxygen functional groups of nanoporous carbons on adsorption of dibenzothiophenes. *Carbon*, **49**, pp. 1216–1224.

57. Seredych, M., Wu, C. T., Brender, P., Ania, C. O., Vix-Guterl, C., and Bandosz, T. J. (2012). Role of phosphorus in carbon matrix in desulfurization of diesel fuel using adsorption process. *Fuel*, **92**, pp. 318–326.

58. Seredych, M., Khine, M., and Bandosz, T. J. (2010). Enhancement in dibenzothiophene reactive adsorption from liquid fuel *via* incorporation of sulfur heteroatoms into the nanoporous carbon matrix. *Chem. Sus. Chem.*, **17**, pp. 137–149.

59. Puziy, A. M., Poddubnaya, O. I., Martýnez-Alonso, A., Suarez-Garcýa, F., and Tascon, J. M. D. (2002). Synthetic carbons activated with phosphoric acid: I. Surface chemistry and ion binding properties. *Carbon*, **40**, pp. 1493–1505.

60. Puziy, A. M., Poddubnaya, O. I., Ziatdinov, A. M. (2006). On the chemical structure of phosphorus compounds in phosphoric acid-activated carbon. *Appl. Surf. Sci.*, **252**, pp. 8036–8038.

61. Puziy, A. M., Poddubnaya, O. I., Sochab, R. P., Gurgulb, J., and Wisniewskic, M. (2008). XPS and NMR studies of phosphoric acid activated carbons. *Carbon*, **46**, pp. 2113–2123.

62. Lee, Y. J., and Radovic. L. R. (2003). Oxidation inhibition effects of phosphorus and boron in different carbon fabrics. *Carbon*, **41**, pp. 1987–1997.

63. Wu, X., and Radovic, L. R. (2006). Inhibition of catalytic oxidation of carbon/ carbon composites by phosphorus. *Carbon*, **44**, pp. 141–51.

64. Rusop, M., Mominuzzaman, S. M., Soga, T., Jimbo, T., and Umeno, M. (2003). Characterization of phosphorus-doped amorphous carbon and construction of n-carbon/p-silicon heterojunction solar cells. *Jpn. J. Appl. Phys.*, **42**, pp. 2339–2344.

65. Mahfoud, K., Loghmarti, M., Muller, J. C., and Siffert, P. (1996). Influence of carbon and oxygen on phosphorus and aluminium co-gettering in silicon solar cells, *Mater. Sci. Eng. B*, **36**, pp. 63–67.

66. Imamura, R., Matsui, K., Takeda, S., Ozaki, J., and Oya, A. (1999). A new role for phosphorus in graphitization of phenolic resin. *Carbon*, **37**, pp. 261–267.

67. Wisniewski, M., Pacholczyk, A., Terzyk, A. P., and Rychlicki, G. (2011). New phosphorus-containing spherical carbon adsorbents as promising materials in drug adsorption and release. *J. Colloid Interf. Sci.*, **354**, pp. 891–894.

68. Pinchin, F. J. (1958). The chlorination of a medium-rank coal at various temperatures. *Fuel*, **37**, pp. 293–298.

69. Yang, C. M., and Kaneko, K. (2002). Adsorption properties of iodine-doped activated carbon fiber. *J. Colloid Interf. Sci.*, **246**, pp. 34–39.

70. Park, S. J., Seo, M. K., and Lee, Y. S. (2003). Surface characteristics of fluorine-modified PAN-based carbon fibers. *Carbon*, **41**, pp. 723–730.

71. Jones, L. E., and Thrower, P. A. (1990). The influence of structure on substitutional doping: SIMS analysis of boron doped pyrolytic graphites. *Carbon*, **28**, pp. 239–241.

72. Seghi, S., Fabio, B., and Economy, J. (2004). Carbon/carbon-boron nitride composites with improved wear resistance compared to carbon/carbon. *Carbon*, **42**, pp. 3043–3048.

73. Wang, D. W., Li, F., Chen, Z. G., Lu, G. Q., and Cheng, H. M. (2008). Synthesis and electrochemical property of boron-doped mesoporous carbon in supercapacitor. *Chem. Mater.*, **20**, pp. 7195–7200.

74. Wu, H. Y., Fan, X. F., Kuo, J. K., and Deng, W. Q. (2010). Carbon doped boron nitride cages as competitive candidates for hydrogen storage materials. *Chem. Commun.*, **46**, pp. 883–885.

75. Bansal, R. C., Dhami, T. L., and Parkash, S. (1977). Surface characteristics and surface behavior of polymer carbons—I: associated oxygen and hydrogen. *Carbon*, **15**, pp. 157–160.

76. Toebes, M. L., Van Dillen, J. A., and de Jong, K. P. (2001). Synthesis of supported palladium catalysts. *J. Mol. Catal. A*, **173**, pp. 75–98.

77. Fraga, M. A., Jordao, E., Mendes, M. J., Freitas, M. M. A., Faria, J. L., and Figueiredo, J. L. (2002). Properties of carbon-supported platinum catalysts: role of carbon surface sites. *J. Catal.*, **209**, pp. 355–364.

78. Sepulveda-Escribano, J. A., Coloma, F., and Rodrýguez-Reinoso, F. (1998). Platinum catalysts supported on carbon blacks with different surface chemical properties. *Appl. Catal. A*, **173**, pp. 247–257.

79. Sun, X. M., and Li, Y. D. (2004). Colloidal carbon spheres and their core/shell structures with noble-metal nanoparticles, *Angew. Chem. Int. Ed.*, **43**, pp. 597–601.

80. Matos, J., Garcia, A., and Titirici M. M. (2010). Solvothermal carbon-doped TiO_2 photocatalyst for the enhanced methylene blue degradation under visible light. *Appl. Catal. A: General*, **390**, pp. 175–182.

81. Velasco, L. F., Dentzer, J., Gadiou, G., Parmentier, J., Vix-Guterl, C., and Ania, C. O. (2011). Synthesis of TiO_2/core shell composites for the photodegradation of organic compounds. *Proceedings of the Carbon International Conference 2011*, **143**, Shanghai.

82. Buczek, B., Biniak, S., and Swiatkowski, A. (1999). Oxygen distribution within oxidized active carbon granules. *Fuel*, **78**, pp. 1443–1448.

83. Ania, C. O., Parra, J. B., Arenillas, A., Rubiera, F., Bandosz, T. J., and Pis, J. J. (2007). On the mechanism of reactive adsorption of dibenzothiophene on organic waste derived carbons. *Appl. Surf. Sci.*, **253**, pp. 5899–5903.

84. Boudou, J. P., Parent, Ph., Suarez-Garcia, F., Villar-Rodil, S., Martinez-Alonso, A., and Tascon, J. M. D. (2006). Nitrogen in aramid-based activated carbon fibers by TPD, XPS and XANES. *Carbon*, **44**, pp. 2452–2462.

85. Kyotani, T., Karasawa, S., Tomita, A. (1986). A TPD study of coal chars in relation to the catalysis of mineral matter. *Fuel*, **65**, pp. 1466–1469.

86. Jansen, R. J. J., and Bekkum, H. (1995). XPS of nitrogen-containing functional groups on activated carbon. *Carbon*, **33**, pp. 1201–1207.

87. Terzyk, A. P. (2001). The influence of activated carbon surface chemical composition on the adsorption of acetaminophen (paracetamol) *in vitro*: Part II. TG, FTIR, and XPS analysis of carbons and the temperature dependence of adsorption kinetics at the neutral pH. *Colloids Surf. A: Physic. Eng. Aspects*, **177**, pp. 23–45.

Terzyk, A. P. (2003). Further insights into the role of carbon surface functionalities in the mechanism of phenol adsorption. *J. Colloids Interf. Sci.*, **268**, pp. 301–3295.

88. Freitas, J. C. C., Emmerich, F. G., Cernicchiaro, G. R. C., Sampaio, L. C., and Bonagamb, T. J. (2001). Magnetic susceptibility effects on 13C MAS NMR spectra of carbon materials and graphite. *Solid State Nucl. Magn. Reson.*, **20**, pp. 61–73.

89. Haghseresht, F., Lu, G. Q., and Whittaker, A. K. (1999). Carbon structure and porosity of carbonaceous adsorbents in relation to their adsorption properties. *Carbon*, **37**, pp. 1491–1497.

90. Tsiao, C., and Botto, R. E. (1991). Xe-129 NMR investigation of coal micropores. *Energy Fuels*, **5**, pp. 87–92.

91. Simonov, P. A., Filimonova, N. S. V., Kruykova, G. N., Boehm, H. P., Moroz, E. M., Likholobov, V. A., and Kuretzky, T. (1999). ^{129}Xe NMR study of carbonaceous materials: effects of surface chemistry and nanotexture. *Carbon*, **37**, pp. 591–600.

92. Kinoshita, K. (1988). *Carbon: Electrochemical and Physicochemical Properties* (JohnWiley & Sons, New York) pp. 293–387.

93. Biniak, S., Swiatkowski, A., and Pakula, M. (2001). *Chemistry and Physics of Carbon*, **27**, ed. Radovic, L. R., "Electrochemical studies of phenomena at active carbon-electrolyte solution interfaces" (Dekker, New York) pp. 125–225.

94. Ania, C. O., and Beguin, F. (2007). Electrochemically assisted adsorption/desorption of bentazone on activated carbon cloth. *Adsorption*, **13**, pp. 579–586.

95. Ania, C. O., and Beguin, F. (2007). Mechanism of adsorption and electrosorption of bentazone on activated carbon cloth in aqueous solutions. *Wat. Res.*, **41**, pp. 3372–3380.

96. Leon-Leon, C. A., and Radovic, L. R. (1990). *Chemistry and Physics of Carbon*, **24**, ed. Thrower, P. A., "Interfacial chemistry and electrochemistry of carbon surfaces" (Marcel Dekker, New York) pp. 212–310.

97. Frackowiak, E., Lota, G., Machnikowski, J., Vix-Guterl, C., and Béguin F. (2006). Optimisation of supercapacitors using carbons with controlled nanotexture and nitrogen content. *Electrochim. Acta*, **51**, pp. 2209–2214.

98. Li, W., Chen, D., Li, Z., Shi, Y., Wan, Y., Huang, J., Yang, J., Zhao, D., and Jiang, Z. (2007). Nitrogen enriched mesoporous carbon spheres obtained by a facile method and its application for electrochemical capacitor. *Electrochem. Commun.*, **9**, pp. 569–573.

99. Haro, M., Rasines, G., Macías, C., and Ania, C. O. (2011). Stability of a carbon gel electrode when used for the electro-assisted removal of ions from brackish water. *Carbon*, **49**, pp. 3753–3730.

100. Berenguer, R., Marco-Lozar, J. P., Quijada, C., Cazorla-Amoros, D., and Morallon, E. (2009). Effect of electrochemical treatments on the surface chemistry of activated carbon. *Carbon*, **47**, pp. 1018–1027.

101. Kinoshita, K., and Bett, J. A. S. (1973). Potentiodynamic analysis of surface oxides on carbon blacks. *Carbon*, **11**, pp. 403–411.

102. Fanning, P. E., and Vannice, M. A. (1991). A DRIFTS study of the formation of surface groups on carbon by oxidation. *Carbon*, **31**, pp. 721–730.

103. Socrates, G. (1994). *Infrared Characteristic Group Frequencies* (Wiley & Sons, Chichester, England).

104. Pels, J. R., Kapteijn, F., Moulijn, J. A., Zhu, Q., and Thomas, K. M. (1995). Evolution of nitrogen functionalities in carbonaceous materials during pyrolysis. *Carbon*, **33**, pp. 1641–1653.

105. Lazar, G., and Lazar, I. (2003). IR characterization of a-C:H:N films sputtered in Ar/CH_4/N_2 plasma. *J. Non-Crystalline Solids*, **331**, pp. 70–78.

106. Boehm, H. P. (1960), ed. Eley, D. D., Pines, H., and Weisz, P. B., *Adv. Catal.*, **16**, "Chemical identification of surface groups" (Academic Press, New York), p. 179.

107. Adib, F., Bagreev, A., and Bandosz, T. J. (2000). Adsorption/oxidation of hydrogen sulfide on nitrogen containing activated carbons. *Langmuir*, **16**, pp. 1980–1986.

108. Jagiello, J., Bandosz, T. J., Putyera, K., and Schwarz, J. A. (1995). Determination of proton affinity distributions for chemical systems in aqueous environments using a stable numerical solution of the adsorption integral equation. *J. Colloid Interf. Sci.*, **172**, pp. 341–346.

109. Ania, C. O., and Bandosz, T. J. (2005). Importance of structural and chemical heterogeneity of activated carbon surfaces for adsorption of dibenzothiophene. *Langmuir*, **21**, pp. 7752–7759.

110. Rudzinski, W., Jagiello, J., and Grillet, Y. (1982). Physical adsorption of gases on heterogeneous solid surfaces: evaluation of adsorption energy distribution from adsorption isotherms and heat of adsorption. *J. Colloid Interf. Sci.*, **87**, pp. 478–491.

111. Jagiello, J. (1994). Stable numerical solution of the adsorption integral equation using splines. *Langmuir*, **10**, pp. 2778–2785.

112. Contescu, A., Contescu, C., Putyera, K., and Schwarz, J. A. (1997). Surface acidity of carbons characterized by their continuous pK distribution and Boehm titration. *Carbon*, **35**, pp. 83–94.

113. ASTM International (1999). Standard test method for pH of activated carbon. *ASTM* D3838–D3880.

114. Noh, J. S., and Schwarz, J. A. (1989). Estimation of the point of zero charge of simple oxides by mass titration. *J. Colloid Interface Sci.*, **130**, pp. 157–164.

115. Ania, C. O., Cabal, B., Parra, J. B., and Pis, J. J. (2007). Importance of the hydrophobic character of activated carbons on the removal of naphthalene from the aqueous phase. *Ads. Sci. Technol.*, **25**, pp. 155–168.

Chapter 2

Preparation of Carbon Materials from Lignocellulosic Biomass

Cesar Nieto Delgado and Jose Rene Rangel Mendez

Institute for Scientific and Technological Research of San Luis Potosí, Environmental Sciences Division, San Luis Potosi, S.L.P., Mexico

rene@ipicyt.edu.mx

2.1 Introduction

Activated carbon is one of the most versatile materials. It has applications in very diverse areas such as catalysis, electrochemistry, electronics, and adsorption processes in both aqueous and gas phase. The success of activated carbon among these areas is mainly due to its extensive surface area and diverse chemical functionalities that can be modified during and after its production process. In recent decades, millions of tons of activated carbon have been produced by the carbon industry, consuming tons of precursors such as mineral carbon and wood. Nowadays, the production of activated carbon from waste biomass is encouraging since by processing organic wastes it is possible to generate products of high commercial value.

Green Carbon Materials: Advances and Applications
Edited by Thomas E. Rufford, Denisa Hulicova-Jurcakova, John Zhu

ISBN 978-981-4411-13-4 (Hardcover), 978-981-4411-14-1 (eBook)
www.panstanford.com

The rate of human population growth in 2010 was 90 million per year, increasing the demand of diverse products to cover their living demands [1]. Agriculture and the agro industries are the responsible to sustain most of these demands. According with the Food and Agriculture Organization of the United Nations (FAO), the world production of sugar cane, maize, rice and wheat in 2008 reached 3.9×10^9 metric tons, generating similar or even higher amounts of agro wastes, such as sugar cane bagasse, corn cobs, straw, and husk of rice and wheat [2]. In most of the cases these, agro wastes have little or no economic value, causing disposal problems. Therefore, using agro wastes to produce carbon materials would reduce the cost of waste disposal and also provide a potentially inexpensive alternative to the existing commercial activated carbons.

2.2 Chemistry of Lignocellulosic Materials

The remaining biomass of a processed crop is mainly composed of tissues that form part of the plants cells structure. The difference between plant cells and animal cells is that plant cells have a cell wall. This wall provides rigidity and mechanical support to the plant cell, maintaining the shape and the direction of cell growth. The cell wall is mainly composed of three biopolymers: cellulose, hemicelluloses, and lignin. The fraction of these main components varies among species, with age, growth season, and with the natural recourses available during the synthesis [3]. Table 2.1 shows the elemental composition and the fiber analysis of diverse lignocellulosic wastes that have been used as activated carbon precursor.

Moreover, the cell wall of a plant is composed of various layers: primary wall, secondary wall, and middle lamella (see Fig. 2.1), which are differentiated by their chemical composition. The primary cell wall is mostly made of cellulose molecules that are unbranched polymers consisting of D-glucose molecules, which are connected to each other by glycosidic (β 1–4) linkages (Fig. 2.2a). Each glucose unit is rotated by 180° from its neighbor performing very long straight chains. This polymer is about 0.8 nm in its maximum width and 0.33 nm^2 in cross-sectional area and can contain about 10,000 glucose residues with their rings in the same plane. These chains are associated with one another to form cellulose crystallites.

Four of these basic crystalline aggregates are then held together by a monolayer of hemicelluloses, generating 25 nm-wide thread-like structures that are enclosed in a matrix of hemicellulose and protolignin (see Fig. 2.1).

Table 2.1 Chemical composition and fiber analysis of waste materials employed to produce activated carbon

	Carbon (%)	Cellulose (%)	Hemicellulose (%)	Lignin (%)
Coconut shell [4]	52.8	33.6	25.0	38.4
Coconut husk [4]	51.5	38.0	28.0	32.8
Hardwood [5]	—	43–47	25–35	16–24
Softwood [5]	—	40–44	25–29	25–31
Eucalyptus wood [6, 7]	47.1	57.3	16.8	25.9
Pine wood [4]	48.0	51.4	24.5	19.0
Beech wood [8, 9]	49.6	45.8	31.8	21.9
Palm shell [10]	50.0	29.7	16.9	53.4
Walnut shell [11, 12]	45.1	25.6	22.1	52.3
Apricot stone [13, 14]	52.4	39.8	34.5	25.8
Peach stone [13, 15]	51.4	26.6	24.3	47.7
Olive stones [16]	46.0	14.0	15.0	42.0
Plum stones [16]	50.7	23.0	20.0	49.0
Agave Bagasse [17, 18]	45.0	47.3	12.8	10.1

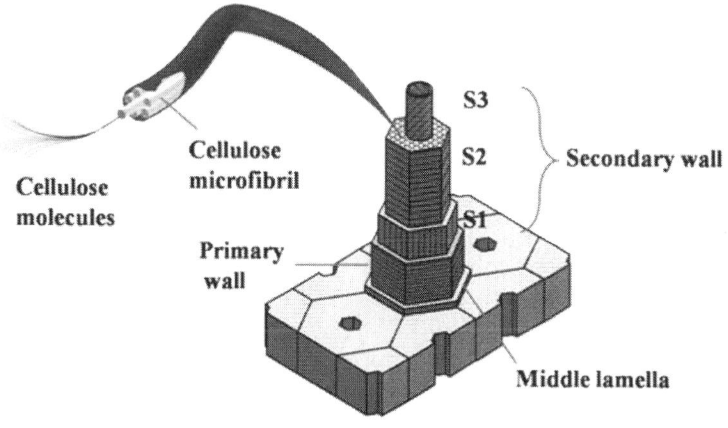

Figure 2.1 Schematic representation of cell wall structure and its main component, the cellulose microfibril.

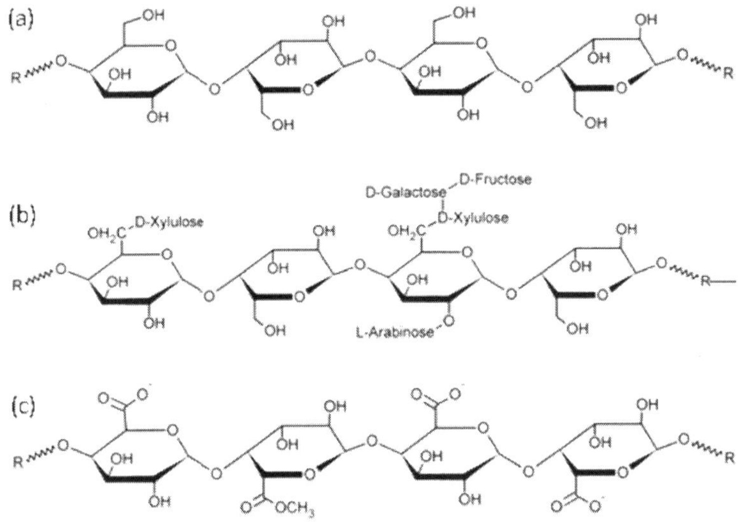

Figure 2.2 Chemical formula of the major components of lignocellulosic materials.

The natural composite that results from this close association is referred to as cellulose microfibril [19, 20]. These microfibrils are the basic unit of the primary and the secondary cell wall, tending to be interwoven in the primary cell wall and parallel to each other in the secondary cell wall [21].

As previously mentioned, the hemicelluloses play an important role in the structure of the plant cell wall. This heteropolymer acts as a binder between cellulose microcrystals to perform the cellulose microfibrils. The hemicellulose chemical structure consists of a variety of polysaccharides that contain, in addition to D-glucose, other carbohydrates such as the hexoses D-mannose, D-galactose, D-fucose, and the pentoses D-xylose and L-arabinose (Fig. 2.2b).

Pectin has a similar function to hemicelluloses, since this also acts as a binder between cellulose microfibrils in the primary cell wall. Pectin is a complex mixture of polysaccharides that consists primarily of 1,4-linked a-D-galacturonic acid residues (Fig. 2.2c): the carboxyl groups of which are normally dissociated and have a negative charge, which gives an ion exchange capacity to the cell wall [21].

Continuing with the plants cell walls composition, the secondary cell wall is located between the plasma membrane and the primary cell wall. This structure incorporates lignin to its chemical composition, which is a phenolic macromolecule with a variable chemical structure. Most of the authors base the chemical structure of lignin on three monomeric precursors: coniferyl alcohol, sinapyl alcohol, and p-coumaryl alcohol, which interconnect to form a polyaromatic structure (Fig. 2.3).

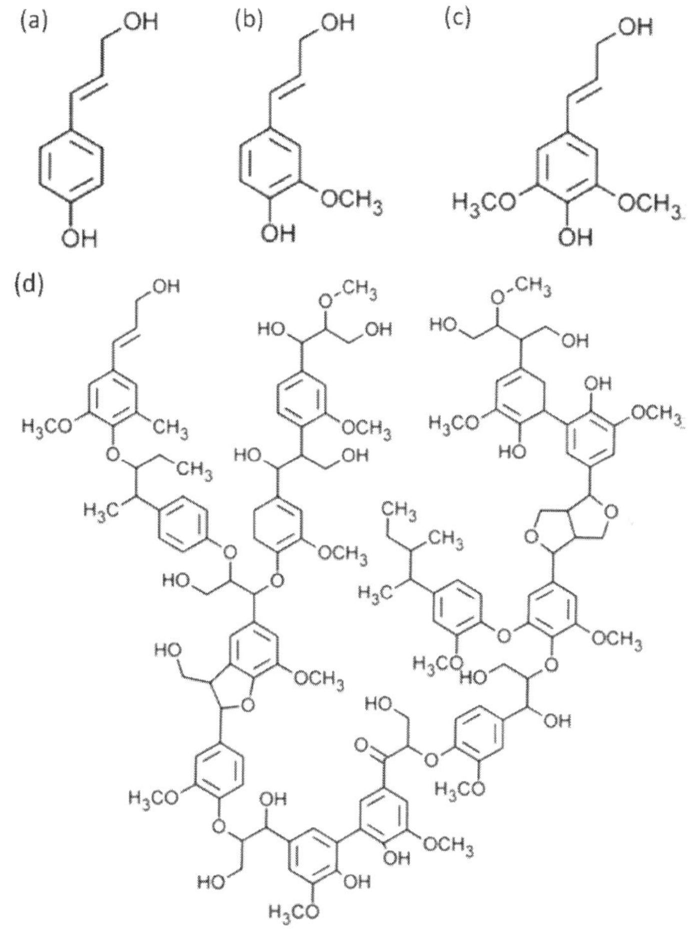

Figure 2.3 Chemical structure of the lignin precursors: (a) p-coumaryl alcohol, (b) coniferyl alcohol, and (c) sinapyl alcohol. (d) The representative chemical structure of lignin (d).

In the secondary cell wall, the cellulose microfibrils and hemicelluloses chains are immersed in a complex mixture of lignin, generating a tighter and rigid structure. Cellulose and hemicellulose appear to be more structurally organized in the secondary cell wall, generating three different layers referred as S1, S2, and S3 in Fig. 2.1, each one with a different arrangement of microfibrils [22, 23]. Finally, the extractives are a minor component of the lignocellulosic materials. These are organic substances that have low molecular weight and are soluble in neutral solvents. Extractives can be composed mainly by resin, fats, waxes, fatty acids, alcohols, terpentines, tannins, and flavonoids [19].

2.3 Activation Processes

The physical and chemical properties of an activated carbon depend on both the precursor characteristics and the activation process employed. In this sense, by controlling the activation process (physical or chemical), it is possible to generate a wide variety of activated carbons to satisfy the requirements of diverse applications. The physical activation process involves the carbonization of the raw material under inert atmosphere, followed by activation at higher temperature in the presence of oxidizing gases. On the other hand, the chemical activation process involves carbonization and activation simultaneously by means of a catalyst (activating agent) that promotes diverse reactions. The combination of both activating processes has been successfully applied. The following sections contain a more detailed description of the most used activation methodologies to produce activated carbons from lignocellulosic materials.

2.3.1 Chemical Activation

As previously mentioned, the production of activated carbon by chemical activation implies the carbonization of the precursor with a reagent often called activating agent. The chemistry behind the chemical activation process depends of the reagent employed. However, most of the activating agents catalyze reactions that prevent the gasification of carbon and promote the formation of pore structure by eliminating non-carbon elements.

The first step when producing activated carbon by chemical activation is the incorporation of the activating agent on the precursor, which is generally milled and dried to ensure a good contact with the reagent. An easy way to impregnate the precursor is by mixing it with the reagent previously dissolved in water. However, when NaOH or KOH are employed as activating agents, a physical mixture of the precursor with the metal hydroxide has proved to be a good impregnation method [24]. When the reagent is dissolved in water, the mixture precursor-activating agent is stirred for some hours to ensure the hydration of the precursor and a better access of the activating agent to all the natural porosity. Further heating of the mixture allows a partial or complete evaporation of water, getting in this way the impregnated precursor. Something to point out is that during the impregnation, the biopolymers of the lignocellulosic materials start to degrade, mostly by hydrolysis reactions catalyzed by the low pH of the activating agent solution or by the heat during the water evaporation. Once the impregnation step ends, the material carbonization is carried out in an inert gas atmosphere at temperatures around 600°C for activating agents such as $ZnCl_2$ and H_3PO_4 and at temperature up to 900°C for metal hydroxides. Finally, the resulting carbon is water washed to eliminate the remaining activating agent.

Many reagents have been tested as activating agents; the most widely used are $ZnCl_2$, H_3PO_4, KOH, and NaOH. Out of these, $ZnCl_2$ and H_3PO_4 are preferred when activating lignocellulosic materials, but when high-rank coals (carbon content >90%), chars, and petroleum coke are the precursors, alkyl hydroxides are often used as the activating agent. Several studies explain the mechanism by which activating agents influence the formation of pore structure. The following section gives a brief description of the promoted reactions by the activating agents often used in chemical activation: zinc chloride, phosphoric acid and metal alkyl hydroxides.

2.3.1.1 Phosphoric acid activation

H_3PO_4 acid has been used as activating agent of lignocellulosic materials such as wood and some agricultural by-products such as almond shells and olive stones as well as carbonaceous materials with high volatile content [25]. These carbons can reach surface areas of 1500 m^2/g. However, when high-rank carbons are employed as precursor, the surface area does not reach these values [26]. This effect is related to the activation mechanism promoted by phosphoric acid. Marit Jagtoyen, Frank Derbyshire and coworkers have carried out extensive studies of H_3PO_4 activation with different raw materials and have proposed an activation mechanism that can be summarized as follows [27–29]: phosphoric acid promotes an acid hydrolysis of the biopolymers that compose the carbon precursor during the impregnation step. This step is characterized by a decrement in the mechanical resistance of the precursor. The precursor becomes elastic by the separation of the cellulose fibers promoted by the phosphoric acid and also by the partial depolymerization of hemicelluloses and lignin. This effect allows better interaction between the precursor and phosphoric acid and generates carbons with better physical properties. After this, during the thermal treatment, the acid media leads the protonation of the different oxygenated groups on the polysaccharides (Fig. 2.4a). Once the oxygen is protonated, diverse reactions can occur. The main reactions are dehydration (Fig. 2.4b), hydrolysis of glycosidic linkages in polysaccharides (Fig. 2.4c), and degradation of sugar monomers (Fig. 2.4d). Also during this stage, the acid attacks the lignin cleaving the aril ether bonds.

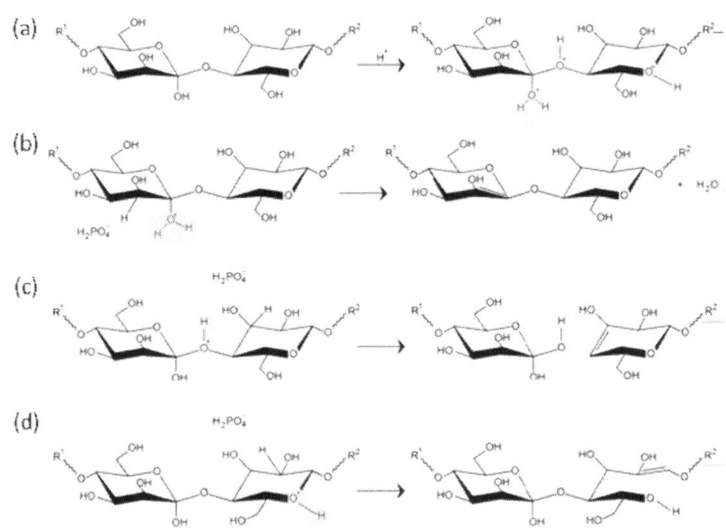

Figure 2.4 Chemical reactions acid catalyzed by H_3PO_4: (a) protonation of different oxygenated groups of sugar monomers. After that hydrogen elimination promotes double bond formation (b), cleavage of the glycosidic linkages between polysaccharides (c), and the degradation of sugar monomers (d).

During the activation at low temperatures (100–200°C), an accelerated weight loss and a volumetric contraction of the precursor have been observed, mainly due to the removal of the volatile products of low molecular weight formed by depolymerization and dehydrating reactions. At higher temperatures, the weight loss rate considerably slows, because H_3PO_4 inhibits the formation of volatile cellulose products as presented in Fig. 2.5a. Also, it was observed that at an activation temperature around 400°C, the precursor begins to dilate. This dilation is related to cross-linking reactions between the polymer chains by the formation of ester linkages between phosphoric acid and OH groups (Fig. 2.5b). As the temperature increases, cyclization and condensation reactions lead to an increase in aromaticity and in the size of the polyaromatic units, enabled by the scission of P–O–C bonds. After the removal of acid on the washing step, the matrix remains dilated by the polyphosphates bridges generated by the activating agent, leaving an accessible pore structure.

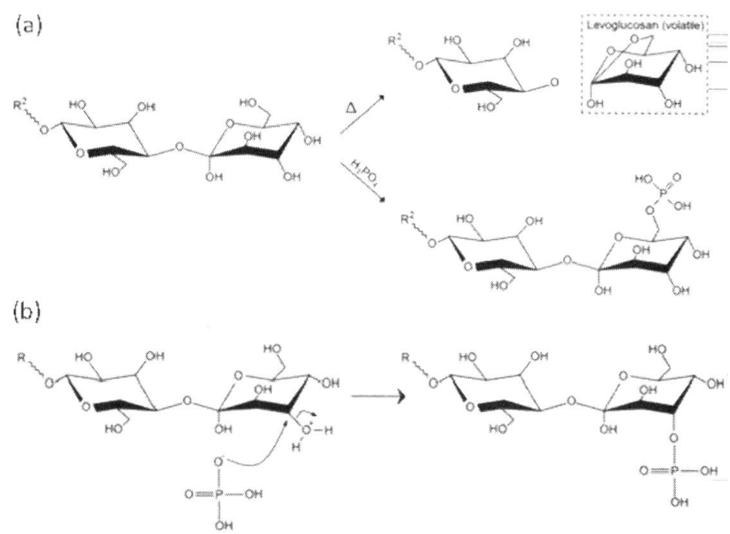

Figure 2.5 (a) Reaction mechanism by which esterification of phosphoric acid would block the formation of cyclic levoglucosan. (b) Formation of ester linkages between phosphoric acid and OH groups of biopolymers.

The phosphorus content in the carbon material considerable rises after activation, giving an interesting mixture of surface functional groups in the activated carbon. According with the reactions that could occur between phosphoric acid and lignocellulosic materials, this activation process requires a heat treatment of about 200°C to dehydrate the precursor and then a further treatment at above 350°C to promote the expansion of the precursor and the consolidation of the pore structure. Hence, the usual activation temperatures employed are between 350 and 550°C (see Table 2.2).

Table 2.2 Effect of activation conditions on the surface area and porous structure of activated carbons produced with H_3PO_4

Precursor	*T* (°C)	*R* (g/g)	Ti (min)	BET (m^2/g)	V_{micro} (cm^3/g)	V_{meso} (cm^3/g)	V_{tot} (cm^3/g)
Olive stones [30]	200	0.50	60	525	0.225	0.035	0.260
	300	0.50	60	1436	0.607	0.146	0.754
	400	0.50	60	1464	0.589	0.267	0.854
	500	0.50	60	1316	0.546	0.230	0.775
	400	0.35	120	1202	0.509	0.094	0.603
	400	0.50	120	1740	0.752	0.139	0.891
	400	0.65	120	1428	0.577	0.331	0.909
Jack fruit peel [31]	350	2.00	45	5	—	—	—
	450	2.00	45	937	0.428	0.135	0.563
	550	2.00	45	1137	0.458	0.215	0.674
	350	3.00	45	4	—	—	—
	450	3.00	45	965	0.431	0.147	0.578
	550	3.00	45	1090	0.471	0.262	0.733
Rice staw [32]	350	1.00	120	551	0.205	0.360	0.570
	425	1.00	120	658	0.250	0.750	0.750
	500	1.00	120	466	0.170	0.390	0.570
	450	0.50	120	571	0.210	0.290	0.500
	450	1.00	120	522	0.180	0.370	0.550
	450	1.60	120	786	0.280	0.770	1.050
Agave bagasse [33]	400	0.80	15	789	0.279	0.051	0.396
	400	1.30	15	1132	0.374	0.265	0.715
	400	3.00	15	998	0.221	0.618	0.953

Abbreviations: T, activation temperature; *R*, activating agent/bagasse weight ratio; Ti, activation time.

The pore structure and the surface area of the activated carbons obtained by phosphoric acid activation are highly influenced by the concentration of phosphoric acid employed. The acid to precursor ratio often employed during the activation of lignocellulosic materials is 1.5 g H_3PO_4/g precursor, generating activated carbons with a wide opened microporous structure and also with an important contribution of mesopores. However, this is highly dependent of the precursor as can be observed in see Table 2.2.

2.3.1.2 Zinc chloride activation

$ZnCl_2$ is often used to activate precursors with high oxygen content, such as low-rank carbons and lignocellulosic materials. This can be attributed to the high reactivity between zinc chloride and the OH groups widely distributed along the main components of the lignocellulosic materials. During the impregnation step, degradation of lignin and cellulose is promoted by the high acidity of the zinc chloride solution. This is because when zinc chloride is dissolved in water, the solution pH decreases due to the protons released during the solvation of zinc ions (Eq. 2.1). This acid catalyzes the degradation even during the evaporation of the activating agent solution.

$$Zn[H_2O]_6^{+2} \rightleftarrows ZnOH\,[H_2O]_5^{+} + H^{+} \qquad (2.1)$$

The early steps of the thermal treatment of $ZnCl_2$-impregnated precursor are always characterized for the release of huge amounts of water. This water evolution has been attributed to dehydrating reactions catalyzed by the activating agent. Zinc chloride can be considered a Lewis acid, which means it is able to interact with the nonbonding electrons of oxygen atoms in the precursor. Most of the oxygen in the biopolymers is found as OH groups attached to the main chain of these or as a linkage between sugar monomers. These functional groups can react with the activating agent by forming Zn–O complexes, follow by the further elimination of water by similar mechanism to those presented in Fig. 2.4. After these reactions, the biopolymers contained in the lignocellulosic material are transformed into randomly oriented carbon structures that generate cavities between them, generating an increase of the surface area. Also, a considerable amount of pores is generated by the dissolution of the remaining zinc chloride, during the rinsing step, after the activation process [34, 35].

The activation temperature often employed when zinc chloride is used as activating agent goes from the zinc chloride melting point, which is around 300–600°C. Generally, higher temperatures can cause the volatilization of high amounts of $ZnCl_2$ and also the decreasing of pore volume and surface area by the heat shrinkage as can be seen in Table 2.3. On the other hand, by varying the amount of $ZnCl_2$ during the activation process, it is possible to modify the surface area and the pore structure of activated carbons. Generally, when using impregnation ratios below 1.5 g $ZnCl_2$/g precursor, the activated carbons produced are essentially microporous. However, when the weight ratio increases, the diameter of pores on carbons becomes wider, generating micro-meso porous materials with a surface area that grows up as the amount of activating agent employed increases. This tendency continues until a weight ratio of about 3 g $ZnCl_2$/g precursor, where a significant decrease of surface area and pore volume is often observed (Table 2.3).

Table 2.3 Effect of activation conditions on the surface area and porous structure of activated carbons produced with $ZnCl_2$

Precursor	T (°C)	R (g/g)	Ti (min)	BET (m²/g)	V_{micro} (cm³/g)	V_{meso} (cm³/g)	V_{tot} (cm³/g)
Macadamia nut shell [36]	500	1.00	60	1718	0.723	—	—
	600	1.00	60	1540	0.649	—	—
	700	1.00	60	1301	0.572	—	—
Pomegranate seeds [37]	600	0.50	60	709	0.240	0.089	0.329
	600	1.00	60	698	0.254	0.066	0.320
	600	2.00	60	979	0.283	0.280	0.563
	800	0.50	60	455	0.211	0.006	0.217
	800	1.00	60	743	0.304	0.072	0.376
	800	2.00	60	824	0.071	0.279	0.350
Cherry stones [38]	400	3.00	120	1472	0.670	0.130	0.800
	500	3.00	120	1566	0.690	0.150	0.840
	600	3.00	120	1345	0.610	0.110	0.720
	800	3.00	120	992	0.490	0.040	0.530
	500	1.00	120	567	0.280	0.020	0.300
	500	2.00	120	1086	0.540	0.050	0.590
	500	3.00	120	1566	0.690	0.150	0.840
	500	4.00	120	1971	0.740	0.450	1.290
Agave bagasse [33]	560	0.80	25	1381	0.476	0.107	0.694
	500	1.20	40	1850	0.676	0.057	0.880
	500	2.00	40	1659	0.402	0.594	1.128

2.3.1.3 Activation with alkali metal hydroxides

This section groups results related to the chemical activation with two reagents: NaOH and KOH. The reactivity and the way in which both reagents activate the carbon precursor are similar, however, it necessary to mention that the activation results for KOH and NaOH are not necessarily equal when treating the same precursor [24, 39–42]. The activation with metal hydroxides is a completely different process with respect to the two processes previously described ($ZnCl_2$ and H_3PO_4). High-rank carbons and chars are the most suitable precursors in metal hydroxide activation. This is because these reagents react directly with the carbon structure generated after the carbonization of the precursor, according to the following reaction:

$$\mathrm{MOH} + \mathrm{C} \rightleftarrows \mathrm{M} + \mathrm{H_2} + \mathrm{M_2CO_3} \tag{2.2}$$

M can be sodium or potassium. This reaction occurs between 360 and 570°C, since in situ XRD studies revealed that the metal hydroxide is only observed up to 400°C, whereas above 600°C the metal carbonate was observed [42]. The consumption of carbon from the char structure initiates, according to Eq. 2.2, the generation of porosity along the precursor. When the activation temperature reaches 800°C, a considerable amount of CO and CO_2 is observed: This has been attributed to the decomposition of the metal carbonate [43]. Both gases can act as gas-phase activating agents, generating reactions similar to those that occur during physical activation (see Section 2.3.2). In addition to these two activation pathways, it has been found that the metal alkyl generated during the alkyl hydroxide decomposition can be intercalated between the carbon structures, expanding the carbon lattice.

The NaOH/KOH activation is usually carried out in an inert atmosphere. Air cannot be used since this would burn up the precursor and consequently react with the metallic alkyl formed. On the other hand, the way in which the metal hydroxide is loaded on the precursor has a high influence on the surface area and the pore structure of activated carbons, being the physical mix of the precursor with the reagent the methodology that generates better results. According to Eq. 2.2, the suitable activation temperatures to produce activated carbon with alkyl metal hydroxides go from 700 to 900°C (see Table 2.4).

Finally, regarding the amount of activating agent employed, a weight ratio between 1 and 4 g reagent/g precursor are often used to obtain activated carbons with surface areas up to 3000 m^2/g as can be seen in Table 2.4. However, something to remark is that at high impregnation ratios the obtained activated carbons are generally powders of low density.

Table 2.4 Effect of activation conditions on the surface area and porous structure of activated carbons produced with metal hydroxides

Precursor/ activating agent	T (°C)	R (g/g)	Ti (min)	BET (m^2/g)	V_{micro} (cm^3/g)	V_{meso} (cm^3/g)	V_{tot} (cm^3/g)
Macadamia nut shell/KOH [36]	600	1.00	60	628	0.290	—	—
	700	1.00	60	1075	0.479	—	—
	800	1.00	60	1169	0.529	—	—
Carbonized Flamboyant pods/ NaOH [44]	700	1.00	90	304	0.161	0.002	0.163
	700	2.00	90	1321	0.692	0.055	0.747
	700	3.00	90	2436	1.312	0.040	1.352
Corn cob/KOH* [45]	600	0.15	60	0.1	—	—	—
	700	0.15	60	548	—	—	0.260
	750	0.15	60	1004	—	—	0.480
	800	0.15	60	1806	—	—	0.870

Abbreviations: T, activation temperature; R, activating agent/precursor weight ratio; Ti, activation time.

*Chemical activation followed by 1 hour of soaking time under CO_2 atmosphere.

2.3.1.4 Activating agent recovery

One of the major concerns in the production of activated carbon by chemical activation is the environmental implications of the activating agent. Most of the papers dealing with the production of activated carbon by chemical activation do not consider this issue. However, the recovery and reuse of the chemicals employed during the chemical activation could reduce the production cost and the expense related to the treatment and disposal of this waste.

In order to recover and reuse the activating agent, this must act as a catalyst during the activation process. In this way, metal alkyl hydroxides cannot be recovered, since during activation these decompose to form metal carbonates or metal oxides. On the other hand, when using phosphoric acid as activating agent, the chemical recovery varies with the production conditions (see Table 2.5). For example, low chemical recoveries are obtained at high activation temperatures.

Table 2.5 Chemical recovery as a function of the production conditions of diverse activated carbons

Precursor	Activating agent	*T* (°C)	*R* (g/g)	Ti (min)	CR (%)
Macadamia nut shell [36]	$ZnCl_2$	500	1.00	60	64.8
		600	1.00	60	4.9
		700	1.00	60	3.3
	KOH	600	1.00	60	93.0
		700	1.00	60	85.6
		800	1.00	60	76.8
Pomegranate seeds [37]	$ZnCl_2$	600	0.50	60	74.4
		600	1.00	60	80.1
		600	2.00	60	81.2
		800	0.50	60	71.3
		800	1.00	60	77.9
		800	2.00	60	77.5
Agave bagasse [17]	H_3PO_4	250	0.40	20	27.9
		250	0.80	20	48.6
		250	1.00	20	45.6
		350	0.40	20	24.6
		350	0.80	20	17.6
		350	1.00	20	33.7
	$ZnCl_2$	250	0.40	20	88.9
		250	0.80	20	87.5
		250	1.00	20	86.7
		350	0.40	20	88.1
		350	0.80	20	87.4
		350	1.00	20	86.2

(Continued)

Table 2.5 *(Continued)*

Precursor	Activating agent	T (°C)	R (g/g)	Ti (min)	CR (%)
Sterculia alata nutshell [46]	$ZnCl_2$	500	1.00	30	72.0
		500	1.00	60	68.0
		500	1.00	120	55.0
		500	1.00	180	48.0
		300	1.00	60	70.0
		400	1.00	60	68.0
		500	1.00	60	62.0
		600	1.00	60	32.0
		500	1.00	60	63.0
		500	2.00	60	77.0
		500	3.00	60	76.0
		500	5.00	60	75.0

Abbreviations: T, activation temperature; R, activating agent/precursor weight ratio; Ti, activation time; CR, chemical recovery.

This is related to the phosphoric acid capability of forming stable linkage with carbon; hence, this remains in the final product. Finally, zinc chloride is the activating agent that can be recovered in a high percentage that goes from 70 to 95%, obtaining the higher chemical recovery at moderate activation conditions (Table 2.5). The latest is possible since $ZnCl_2$ acts as a catalyst during the activation process: The unrecovered fraction can be related to the evaporation of the molten salt.

2.3.2 Thermal Activation

Physical activation, also called thermal activation, is the process that consists of diverse thermal treatments by which the carbon precursor develops porosity and hence extensive surface area. This process applies for lignocellulosic materials and mineral carbons. However, the preferred materials are high-rank carbons, wood, and lignocellulosic materials such as fruit stones and hard shells. The raw material needs some pretreatment prior to its activation: First, the required particle size must be obtained by crushing and sieving. Also, sometimes a washing step with water or acid is needed in order to remove any undesirable material and to reduce the inorganic content.

Generally, physical activation consists of two consecutive steps. The first step consists of carbonating the precursor at moderate temperatures, to produce the char. The second step comprises the char activation, where it is partially gasified with an oxidizing agent, usually CO_2 or steam. Both processes are described in the following section.

2.3.2.1 Carbonization process

The main objective of the carbonization step is to transform the precursor to char, eliminating the non-carbon atoms that form part of the aromatic structure. During the early stages of this thermal treatment, the volatilization of the most labile tissues of the precursor can occur. In general, the gases that first evolve are those coming from labile carbon such as fats, waxes, alkaloids, proteins, phenolics, simple sugars, pectins, gums, resins, terpenes, starches, glycosides, and essential oils. The subsequent mechanism consists of a complex combination of individual pyrolysis of cellulose, hemicelluloses, and lignin. In addition, secondary reactions can result from the cross-reactions of primary pyrolysis products and the original molecules. Owing to the high correlation between the different components of the plant cells wall, it is difficult to define a combined pyrolysis mechanism for all the components, hence most of the proposed mechanisms take into account components as isolated compounds.

Cellulose degradation occurs between 240 and 350°C, giving two main products: char and volatiles. The main route to gasify cellulose is by means of a free radical mechanism that produces anhydrocellulose and levoglucosan (Fig. 2.5). On the other hand, the interaction of the secondary vapors with the remaining solid phase generates most of the char obtained during cellulose pyrolysis [47]. Hemicelluloses are the most labile of the three basic biopolymers of the lignocellulosic materials. This carbohydrate decomposes at temperatures between 180 and 260°C, generating large amounts of volatiles and less char than cellulose [48]. Finally, lignin is the biomass fraction with the highest decomposition temperature since this decomposes between 280 to 500°C. The pyrolysis of this fraction generates phenols *via* the cleavage of ether and carbon–carbon linkages. Lignin is more difficult to dehydrate than cellulose or hemicelluloses; hence, it is the fraction with the highest contribution of char [48].

Microscope observations of carbonized agave bagasse revealed a structure constituted of the most lignified part of the cytoeskeletal (Fig. 2.6). This material has a higher aromaticity and carbon content; however, the surface area is almost negligible. Generally, this material already has a microporous structure comprising void spaces between the disordered short graphitic crystals. Nevertheless, this porous structure is not accessible because the meso- and macroporous structure is blocked by tar and soot that is condensed during carbonization. The controlled removal of these components generates accessible porous structure and ample surface area, which is the main objective of the activation step.

Figure 2.6 Scanning electron micrographs of carbonized agave bagasse fibers under N_2 atmosphere at 300°C.

2.3.2.2 Activation process

The objective of the activation process is to enlarge and to open the incipient pore structure developed during carbonization. Also, during this process, oxygenated groups can be incorporated to the carbon structure, generating a more "active" carbon for specific applications. To accomplish this task, the reactive atmosphere is changed from inert to oxidant by using gases such as oxygen, steam, and carbon dioxide. Depending on the activating agent employed, different reactions can occur:

$$2C + O_2 \rightarrow 2CO \qquad \Delta H = -387\ \text{kJ/mol} \tag{2.3}$$

$$C + O_2 \rightarrow CO_2 \qquad \Delta H = -226\ \text{kJ/mol} \tag{2.4}$$

$$C + H_2O \rightarrow CO + H_2 \qquad \Delta H = 132\ \text{kJ/mol} \tag{2.5}$$

$$C + CO_2 \rightarrow 2CO \qquad \Delta H = 159\ \text{kJ/mol} \tag{2.6}$$

According with the reaction enthalpy, O_2 is the most reactive oxidant and CO_2 the least reactive. Owing to the highly exothermic enthalpy of oxygen reactions, the temperature of the reaction is extremely difficult to control, usually causing mainly the consumption of the external surface of the precursor. Consequently, oxygen activation is scarcely applied. On the other hand, CO_2 and H_2O are an option to activate carbon under moderate oxidation conditions, since the promoted reactions by both activating agents are endothermic and require external heating for maintaining the reactions [49–51]. Therefore, by regulating the activation temperature and residence time, carbon oxidation can be controlled. Finally, other factor that influences the characteristics of the generated activated carbon by thermal activation is the presence of inorganic impurities, because these can catalyze or inhibit the gasification reactions [52–54].

2.3.3 Hydrothermal Carbonization

The thermal treatment of water mixed with organic substances at temperatures between 150 and 350°C (autogenously pressured) is known as hydrothermal carbonization or wet pyrolysis [55]. By this methodology, it is possible to decompose lignocellulosic materials in water-soluble organic substances and in a carbon-rich solid product: This process has generated widespread interest in recent years in diverse research areas [56–59]. As mentioned previously, activated carbon from lignocellulosic precursors is obtained at high temperatures. Most of the reactions causing the carbonization of precursors are accompanied by the evolution of various gases; hence, by modifying the process pressure, it is possible to modify the carbonization reactions and as a consequence the resultant carbon is different in structure and properties of those obtained at ambient pressure. Also, the gases released during activation at normal pressure result in the loss of carbon atoms from the precursor, which means lower carbon yield.

There are numerous reports that explain the reaction mechanisms during thermal carbonization of lignocellulosic materials mostly based on cellulose as model biopolymer [55 ,59–61]. From this information, Sevilla et al. [60] proposed that the hydrothermal carbonization takes place *via* (i) hydrolysis of cellulose chains, (ii) dehydration and fragmentation into soluble products of the monomers that come from the hydrolysis of cellulose, (iii) polymerization or condensation of the soluble products, (iv) aromatization of the polymers thus formed, (v) appearance of a short burst of nucleation, and (vi) growth of the nuclei so formed by diffusion and linkage of species from the solution to the surface of the nuclei. Something to remark is that the reactions promoted during the hydrothermal carbonization have lower activation energy in comparison with the ones that take place during dry pyrolysis. This causes a low degradation temperature of the biomass components: hemicelluloses decompose between 180 and 200°C, most of the lignins between 180 and 220°C and cellulose about 220°C.

Hydrothermal carbonization has been successfully employed to produce activated carbon in a single thermal step [61–65]. The carbon content of the carbons obtained is generally between 75 and 80% with a high oxygen content. Unlike the previous activation method, hydrothermal carbonization degrades most of the initial structure of the precursor, and the carbons obtained consist mainly of aggregates of microspheres with diameters between 5 and 50 μm. These spheres generally have an aromatic core of cross-linked furanic rings, and aldehydic and carboxyl functional groups at the edges of the sphere [65].

2.3.4 Microwave-Assisted Activation

The application of microwave heating for the activation of lignocellulosic materials has been investigated, and very promising results have been obtained [66–69]. The main difference between microwave activation and conventional heating is the way in which the heat is supplied. Thermal treatments are mostly performed by using an external heating source such as an electric furnace. The external heating supply energy to the surface of raw material and then the heat is transferred inside by heat conduction. Therefore, it is difficult to achieve uniform temperature in the material. In contrast, the microwave irradiation supplies energy to the whole material body, converting the energy contained in the microwaves into heat inside the particles by dipole rotation and ionic conduction [69, 70]. This has some advantages since the precursor can absorb microwave energy quickly and the treatment time can be considerably reduced, which in many cases represents a reduction in the energy consumption and also a reduction in inert gas consumption [71].

2.4 Agave Bagasse as a Precursor of Activated Carbon

Agave bagasse is a lignocellulosic waste obtained during the production of Tequila and Mescal in Mexico. This waste comprises the fibers that remain after the agave plants' heads are cooked and milled to extract the sugars. The national production of Tequila and Mescal in 2008 was around 181 million liters [72, 73]. According with this production, the quantity of agave bagasse rose to 350,000 tons on dry basis [74]. The large amounts of bagasse generated in the distilleries combined with the lack of treatment and disposition control cause several environmental problems. Most of this waste is burned under uncontrolled conditions, generating large amounts of ashes and particulate matter, which are spread into the environment. Taking into account the commercial significance of activated carbon worldwide and the large quantity of agave bagasse that is generated in Mexico, the production of activated carbon from agave bagasse is a feasible alternative for producing low-cost activated carbon, which can also contribute to solving environmental problems at distilleries.

The carbon content of agave bagasse is about 45% with an ash content, mainly calcium, of approximately 6.5% [17]. Moreover, the fiber content is distributed as follows: 12.8% of hemicellulose, 47.3% of cellulose, 10.1% of lignin, and 29.8% of extractives [18], which is similar to other activated carbon precursors (see

Table 2.1). As agave bagasse is a lignocellulosic material with low lignin content, the most suitable activation methodology is chemical activation with a dehydrating agent: $ZnCl_2$ and H_3PO_4 were tested as activating agents. Employing the surface response methodology, it was possible to determine the influence of the production conditions on the surface area and hardness of the activated carbons [33]. Figure 2.7 shows the influence of the activation temperature and activating agent to precursor weight ratio on the surface area and hardness of the agave bagasse-based activated carbons. As can be seen, a wide variety of materials can be obtained with both activating agents by varying the production conditions. Particularly, the $ZnCl_2$-activated carbons developed a higher surface area compared with those obtained with H_3PO_4. However, phosphoric acid-activated carbons have better mechanical properties: Their hardness can reach values as high as 90%, which are similar to the hardness of commercial activated carbon. A detailed discussion about this procedure has been published elsewhere [33]. These

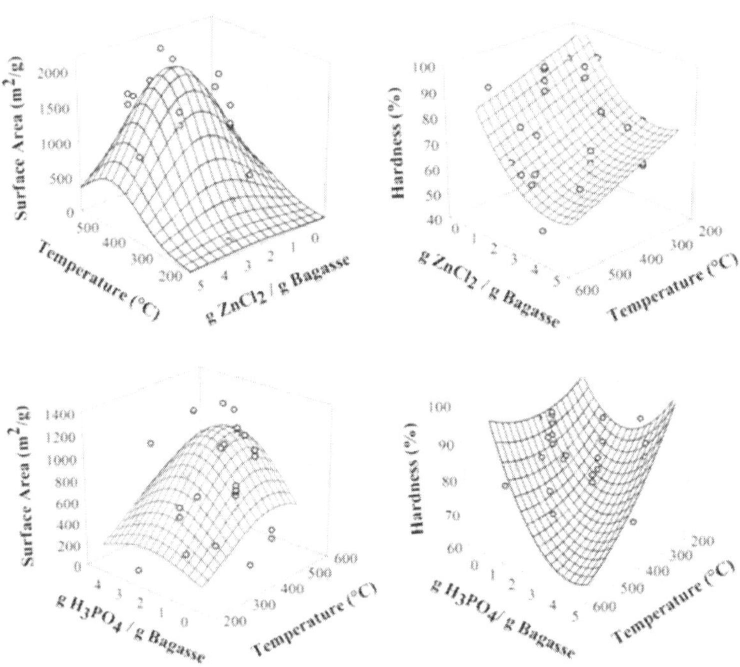

Figure 2.7 Effect of the production conditions on the surface area and hardness of agave bagasse-based activated carbons processed with $ZnCl_2$ and H_3PO_4.

results demonstrated that agave bagasse from the Tequila and Mescal industry, which is an organic waste, can be considered a precursor of activated carbons with reasonable surface area and hardness.

2.5 Conclusions

The use of lignocellulosic materials to produce activated carbon has been of interest to numerous researchers since this represents a "green" alternative for producing the carbonaceous adsorbent most used all over the world. Also, by using waste biomass as an activated carbon precursor, we not only take advantage of the availability and low cost of these materials but also raise an alternative for the treatment and disposal of hundreds of tons of waste biomass generated worldwide, which in most cases leads to a negative environmental impact.

As explained in this chapter, there are a wide variety of biomass and techniques by which the lignocellulosic wastes can be transformed into good-quality carbonaceous materials as has been confirmed by many researchers. However, since lignocellulosic materials have different chemical and botanical structures, the identification of the most adequate methodology to produce activated carbon is needed.

Acknowledgments

We would like to thank FOMIX-SLP (FMSLP-2008-C02-99664) and CIENCIA BASICA (SEP-CB-2008-01-105920) for the financial support to conduct this research; CONACYT for the doctorate fellowship number 204158; and Dulce Partida, Guillermo Vidriales, Carmen Rocha, Gladis Labrada and Beatriz Rivera for their technical support. Finally, we acknowledge LINAN and LANBAMA national laboratories for their support in the samples analyses.

References

1. U. S. Census Bureau. (2010). U. S. Census.
2. Food and Agriculture Organization of the United Nations (2010). *Agricultural Production in FAO Statistical Yearbook 2010.*

3. Brooker, R. J., Widmaier E. P., Graham, L. E., and Stiling, P. D. (2008). *Biology* (McGraw-Hill, New York).
4. Pollard, S. J. T., Fowler, G. D., Sollars, C. J., and Perry, R. (1992). Low-cost adsorbents for waste and wastewater treatment: a review. *The Science of the Total Environment.*, **116**, pp. 31–52.
5. Hon, D. N. S. (1996). *Chemical Modification of Lignocellulosic Materials* (Marcel Dekker, New York).
6. Amaya, A., Pariz, J., Tancredi, N., and Cordero, T. (2007). Activated carbon pellets from eucalyptus char and tar TG studies. *J. Thermal Anal. Calorim.*, **89**, pp. 987–991.
7. Tancredi, N., Cordero, T., Rodríguez-Mirasol, J., and Rodríguez, J. J. (1996). CO_2 gasification of eucalyptus wood chars. *Fuel*, **75**, pp. 1505–1508.
8. Rath, J., Steiner, G., Wolfinger, M. G., and Staudinger, G. (2002). Tar cracking from fast pyrolysis of large beech wood particles. *J. Anal. Appl. Pyrol.*, **62**, pp. 83–92.
9. Demirbas, A. (2003). Relationships between lignin contents and fixed carbon contents of biomass samples. *Energy Conversion Manag.*, **44**, pp. 1481–1486.
10. Adinata, D., Wan Daud, W. M. A., and Aroua, M. K. (2007). Production of carbon molecular sieves from palm shell based activated carbon by pore sizes modification with benzene for methane selective separation. *Fuel Process. Technol.*, **88**, pp. 599–605.
11. González, J. F., Román, S., Encinar, J. M., and Martínez, G. (2009). Pyrolysis of various biomass residues and char utilization for the production of activated carbons. *J. Anal. Appl. Pyrol.*, **85**, pp. 134–141.
12. Kambarova, G., and Sarymsakov, S. (2008). Preparation of activated charcoal from walnut shells. *Solid Fuel Chem.*, **42**, pp. 183–186.
13. Kaynak, B., Topal, H., and Atimtay, A. T. (2005). Peach and apricot stone combustion in a bubbling fluidized bed. *Fuel Process. Technol.*, **86**, pp. 1175–1193.
14. Soleimani, M., and Kaghazchi, T. (2007). Agricultural waste conversion to activated carbon by chemical activation with phosphoric acid. *Chem. Eng. Technol.*, **30**, pp. 649–654.
15. Muñoz-González, Y., Arriagada-Acuña, R., Soto-Garrido, G., and García-Lovera, R. (2009). Activated carbons from peach stones and pine sawdust by phosphoric acid activation used in clarification and decolorization processes. *J. Chem. Technol. Biotechnol.*, **84**, pp. 39–47.

16. Cagnon, B., Py, X., Guillot, A., Stoeckli, F., and Chambat, G. (2009). Contributions of hemicellulose, cellulose and lignin to the mass and the porous properties of chars and steam activated carbons from various lignocellulosic precursors. *Bioresour. Technol.*, **100**, pp. 292–298.

17. Nieto-Delgado, C., Terrones, M., and Rangel-Mendez, J. R. (2011). Development of highly microporous activated carbon from the alcoholic beverage industry organic by-products. *Biomass Bioenergy*, **35**, pp. 103–112.

18. Garcia-Reyes, R. B., and Rangel-Mendez, J. R. (2009). Contribution of agro-waste material main components (hemicelluloses, cellulose, and lignin) to the removal of chromium (III) from aqueous solution. *J. Chem. Technol. Biotechnol.*, **84**, pp. 1533–1538.

19. Nimz, H. (1984). Wood-chemistry, ultrastructure, reactions. *Eur. J. Wood Wood Prod.*, **42**, pp. 314–314.

20. Brett, C. T. (2000). Cellulose microfibrils in plants: Biosynthesis, deposition, and integration into the cell wall. *International Review of Cytology.*, **199**, pp. 161–199.

21. Nobel, P. (2009). *Physicochemical and Environmental Plant Physiology* (Elsevier Academic Press, U.S.).

22. Salisbury, F. B., and Ross, C. W. (1992). *Plant Physiology*. Wadsworth Pub. Co., Belmont, Californiaa.

23. Goodwin, T. W., and Mercer, E. I. (1983). *Introduction to Plant Biochemistry*. Butterworth-Heinemann Ltd, UK.

24. Lillo-Ródenas, M. A., Lozano-Castelló, D., Cazorla-Amorós, D., and Linares-Solano, A. (2001). Preparation of activated carbons from Spanish anthracite: II. Activation by NaOH. *Carbon*, **39**, pp. 751–759.

25. Dias, J. M., Alvim-Ferraz, M. C. M., Almeida, M. F., Rivera-Utrilla, J., and Sánchez-Polo, M. (2007). Waste materials for activated carbon preparation and its use in aqueous-phase treatment: a review. *J. Environ. Manage.*, **85**, pp. 833–846.

26. Villar-Rodil, S., Denoyel, R., Rouquerol, J., Martínez-Alonso, A., and Tascón, J. M. D. (2002). Porous texture evolution in nomex-derived activated carbon fibers. *J. Colloid Interface Sci.*, **252**, pp. 169–176.

27. Jagtoyen, M., and Derbyshire, F. (1993). Some considerations of the origins of porosity in carbons from chemically activated wood. *Carbon*, **31**, pp. 1185–1192.

28. Jagtoyen, M., Thwaites, M., Stencel, J., McEnaney, B., and Derbyshire, F. (1992). Adsorbent carbon synthesis from coals by phosphoric acid activation. *Carbon*, **30**, pp. 1089–1096.

29. Solum, M. S., Pugmire, R. J., Jagtoyen, M., and Derbyshire, F. (1995). Evolution of carbon structure in chemically activated wood. *Carbon*, **33**, pp. 1247–1254.

30. Yavuz, R., Akyildiz, H., Karatepe, N., and Çetinkaya, E. (2010). Influence of preparation conditions on porous structures of olive stone activated by H_3PO_4. *Fuel Process. Technol.*, **91**, pp. 80–87.

31. Prahas, D., Kartika, Y., Indraswati, N., and Ismadji, S. (2008). Activated carbon from jackfruit peel waste by H_3PO_4 chemical activation: pore structure and surface chemistry characterization. *Chem. Eng. J.*, **140**, pp. 32–42.

32. Fierro, V., Muñiz, G., Basta, A. H., El-Saied, H., and Celzard, A. (2010). Rice straw as precursor of activated carbons: Activation with ortho-phosphoric acid. *J. Hazard. Mater.*, **181**, pp. 27–34.

33. Nieto-Delgado, C., and Rangel-Mendez, J. R. (2011). Production of activated carbon from organic by-products from the alcoholic beverage industry: surface area and hardness optimization by using the response surface methodology. *Ind. Crops Prod.*, **34**, pp. 1528–1537.

34. Caturla, F., Molina-Sabio, M., and Rodríguez-Reinoso, F. (1991). Preparation of activated carbon by chemical activation with $ZnCl_2$. *Carbon*, **29**, pp. 999–1007.

35. Molina-Sabio, M., and Rodríguez-Reinoso, F. (2004). Role of chemical activation in the development of carbon porosity. *Colloids Surf. A: Physicochem. Eng. Aspects*, **241**, pp. 15–25.

36. Ahmadpour, A., and Do, D. D. (1997). The preparation of activated carbon from macadamia nutshell by chemical activation. *Carbon*, **35**, pp. 1723–1732.

37. Uçar, S., Erdem, M., Tay, T., and Karagöz, S. (2009). Preparation and characterization of activated carbon produced from pomegranate seeds by $ZnCl_2$ activation. *Appl. Surf. Sci.*, **255**, pp. 8890–8896.

38. Olivares-Marín, M., Fernández-González, C., Macías-García, A., and Gómez-Serrano, V. (2006). Preparation of activated carbon from cherry stones by chemical activation with $ZnCl_2$. *Appl. Surf. Sci.*, **252**, pp. 5967–5971.

39. Lozano-Castelló, D., Lillo-Ródenas, M. A., Cazorla-Amorós, D., and Linares-Solano, A. (2001). Preparation of activated carbons from Spanish anthracite: I. activation by KOH. *Carbon*, **39**, pp. 741–749.

40. Maciá-Agulló, J. A., Moore, B. C., Cazorla-Amorós, D., and Linares-Solano, A. (2004). Activation of coal tar pitch carbon fibres: Physical activation vs. chemical activation. *Carbon*, **42**, pp. 1367–1370.

41. Raymundo-Piñero, E., Cazorla-Amorós, D., Linares-Solano, A., Delpeux, S., Frackowiak, E., Szostak, K., and Béguin, F. (2002). High surface area carbon nanotubes prepared by chemical activation. *Carbon*, **40**, pp. 1614–1617.

42. Raymundo-Piñero, E., Azaïs, P., Cacciaguerra, T., Cazorla-Amorós, D., Linares-Solano, A., and Béguin, F. (2005). KOH and NaOH activation mechanisms of multiwalled carbon nanotubes with different structural organisation. *Carbon*, **43**, pp. 786–795.

43. Lillo-Ródenas, M. A., Cazorla-Amorós, D., and Linares-Solano, A. (2003). Understanding chemical reactions between carbons and NaOH and KOH: an insight into the chemical activation mechanism. *Carbon*, **41**, pp. 267–275.

44. Vargas, A. M. M., Cazetta, A. L., Garcia, C. A., Moraes, J. C. G., Nogami, E. M., Lenzi, E., Costa, W. F., and Almeida, V. C. (2010). Preparation and characterization of activated carbon from a new raw lignocellulosic material: Flamboyant (Delonix regia) pods. *J. Environ. Manage.*, **92**, pp. 178–184.

45. Tsai, W. T., Chang, C. Y., Wang, S. Y., Chang, C. F., Chien, S. F., and Sun, H. F. (2001). Cleaner production of carbon adsorbents by utilizing agricultural waste corn cob. *Resources, Conservation Recycling*, **32**, pp. 43–53.

46. Mohanty, K., Das, D., and Biswas, M. (2006). Preparation and characterization of activated carbons from *Sterculia alata* nutshell by chemical activation with zinc chloride to remove phenol from wastewater. *Adsorption*, **12**, pp. 119–132.

47. Mohan, D., Pittman, C. U., and Steele, P. H. (2006). Pyrolysis of wood/biomass for bio-oil: a critical review. *Energy Fuels*, **20**, pp. 848–889.

48. Rodríguez-Reinoso, F., Molina-Sabio, M., and González, M. T. (1995). The use of steam and CO_2 as activating agents in the preparation of activated carbons. *Carbon*, **33**, pp. 15–23.

49. Molina-Sabio, M., Gonzalez, M. T., Rodriguez-Reinoso, F., and Sepúlveda-Escribano, A. (1996). Effect of steam and carbon dioxide activation in the micropore size distribution of activated carbon. *Carbon*, **34**, pp. 505–509.

50. Román, S., González, J. F., González-García, C. M., and Zamora, F. (2008). Control of pore development during CO_2 and steam activation of olive stones. *Fuel Process. Technol.*, **89**, pp. 715–720.

51. Samaras, P., Diamadopoulos, E., and Sakellaropoulos, G. P. (1991). The effect of demineralization on lignite activation. *Carbon*, **29**, pp. 1181–1190.

52. Samaras, P., Diamadopoulos, E., and Sakellaropoulos, G. P. (1994). Acid treatment of lignite and its effect on activation. *Carbon*, **32**, pp. 771–776.

53. Samaras, P., Diamadopoulos, E., and Sakellaropoulos, G. P. (1996). The effect of mineral matter and pyrolysis conditions on the gasification of Greek lignite by carbon dioxide. *Fuel*, **75**, pp. 1108–1114.

54. Peterson, A. A., Vogel, F., Lachance, R. P., Froling, M., Antal, J. M. J., and Tester, J. W. (2008). Thermochemical biofuel production in hydrothermal media: A review of sub- and supercritical water technologies. *Energy Environ. Sci.*, **1**, pp. 32–65.

55. Wang, Q., Li, H., Chen, L., and Huang, X. (2001). Monodispersed hard carbon spherules with uniform nanopores. *Carbon*, **39**, pp. 2211–2214.

56. Wang, Q., Li, H., Chen, L., and Huang, X. (2002). Novel spherical microporous carbon as anode material for Li-ion batteries. *Solid State Ionics*. **152–153**, pp. 43–50.

57. Titirici, M. M., Thomas, A., Yu, S.-H., Miller, J.-O., and Antonietti, M. (2007). A direct synthesis of mesoporous carbons with bicontinuous pore morphology from crude plant material by hydrothermal carbonization. *Chem. Mater.*, **19**, pp. 4205–4212.

58. Yao, C., Shin, Y., Wang, L.-Q., Windisch, C. F., Samuels, W. D., Arey, B. W., Wang, C., Risen, W. M., and Exarhos, G. J. (2007). Hydrothermal dehydration of aqueous fructose solutions in a closed system. *J. Phys. Chem. C*, **111**, pp. 15141–15145.

59. Funke, A., and Ziegler, F. (2010). Hydrothermal carbonization of biomass: a summary and discussion of chemical mechanisms for process engineering. *Biofuels, Bioprod. Biorefining*, **4**, pp. 160–177.

60. Sevilla, M., and Fuertes, A. B. (2009). The production of carbon materials by hydrothermal carbonization of cellulose. *Carbon*, **47**, pp. 2281–2289.

61. Hu, B., Wang, K., Wu, L., Yu, S.-H., Antonietti, M., and Titirici, M.-M. (2010). Engineering carbon materials from the hydrothermal carbonization process of biomass. *Adv. Mater.*, **22**, pp. 813–828.

62. Mochidzuki, K., Sato, N., and Sakoda, A. (2005). Production and characterization of carbonaceous adsorbents from biomass wastes by aqueous phase carbonization. *Adsorption*, **11**, pp. 669–673.

63. Titirici, M.-M., Thomas, A., and Antonietti, M. (2007). Aminated hydrophilic ordered mesoporous carbons. *J. Mater. Chem.*, **17**, pp. 3412–3418.

64. Zhao, L., Baccile, N., Gross, S., Zhang, Y., Wei, W., Sun, Y., Antonietti, M., and Titirici, M.-M. (2010). Sustainable nitrogen-doped carbonaceous materials from biomass derivatives. *Carbon*, **48**, pp. 3778–3787.

65. Sun, X., and Li, Y. (2004). Colloidal carbon spheres and their core/shell structures with noble-metal nanoparticles. *Angew. Chem. Int. Ed.*, **43**, pp. 597–601.

66. Yagmur, E., Ozmak, M., and Aktas, Z. (2008). A novel method for production of activated carbon from waste tea by chemical activation with microwave energy. *Fuel*, **87**, pp. 3278–3285.

67. Li, W., Zhang, L.-b., Peng, J.-h., Li, N., and Zhu, X.-Y. (2008). Preparation of high surface area activated carbons from tobacco stems with K_2CO_3 activation using microwave radiation. *Ind. Crops Prod.*, **27**, pp. 341–347.

68. Guo, J., and Lua, A. C. (2000). Preparation of activated carbons from oil-palm-stone chars by microwave-induced carbon dioxide activation. *Carbon*, **38**, pp. 1985–1993.

69. Valente Nabais, J. M., Carrott, P. J. M., Ribeiro Carrott, M. M. L., and Menéndez, J. A. (2004). Preparation and modification of activated carbon fibres by microwave heating. *Carbon*, **42**, pp. 1315–1320.

70. Jones, D. A., Lelyveld, T. P., Mavrofidis, S. D., Kingman, S. W., and Miles, N. J. (2002). Microwave heating applications in environmental engineering—a review. *Resources, Conservation Recycling*, **34**, pp. 75–90.

71. Menéndez, J. A., Menéndez, E. M., Iglesias, M. J., García, A., and Pis, J. J. (1999). Modification of the surface chemistry of active carbons by means of microwave-induced treatments. *Carbon*, **37**, pp. 1115–1121.

72. Consejo nacional de Productores de Maguey Mezcal (2009). Informe anual. Available from http://www.comercam.org/files/INFORME%20ANUAL%202009.pdf.

73. Consejo Regulador del Tequila (2007). Informacion estadistica. Available from http://www.crt.org.mx/EstadisticasCRTweb.

74. Rogelio Aguirre Rivera, J., Charcas Salazar, H., Luis Flores Flores, J. (2001). *El Maguey Mezcalero Potosino* (Universidad Autónoma de San Luis Potosí, Mexico).

Chapter 3

Adsorbed Natural Gas and Hydrogen Storage

Juan Alcañiz-Monge, Dolores Lozano-Castelló, Diego Cazorla-Amorós, and Angel Linares-Solano

Inorganic Chemistry, Alicante University, Alicante, E-03080, Spain

jalcaniz@ua.es

This chapter presents an overview of the research of our group in natural gas and hydrogen storage, in which the behavior of different carbon materials in methane and hydrogen storage has been studied. An in-depth analysis has been done in order to obtain the carbon material with the best methane and hydrogen adsorption capacity by unit volume of adsorbent. The important effect of the micropore volume, micropore size distribution (MPSD), and packing density of the carbon materials in methane and hydrogen adsorption capacity and delivery has been analyzed.

3.1 Introduction

The primary energy consumed in the world in the 20th century is obtained from fossil fuels (81.2% in 2008). Coal and natural gas

Green Carbon Materials: Advances and Applications
Edited by Thomas E. Rufford, Denisa Hulicova-Jurcakova, John Zhu

ISBN 978-981-4411-13-4 (Hardcover), 978-981-4411-14-1 (eBook)
www.panstanford.com

(NG) are the main fossil fuels used in the electricity generation (60%), whereas in the transportation (an important sector that accounts for 27% of the world's total delivered energy) liquid fuels remain to be the main combustibles [1]. The rapid increase in world energy prices, combined with concerns about the environmental consequences of greenhouse gas emissions, has led to renewed interest in alternatives to fossil fuels [2]. Consequently, in the past decades, a lot of research has been done in the development and implantation of technologies based on NG and hydrogen as vehicular fuels.

Fortunately, in the past years, the global NG resources have raised spectacularly with the inclusion of the so-called unconventional NG, which has been reached due to the use of new prospection techniques. In the near term, as world economies begin to recover from the downturn, global demand for NG is expected to rebound, with NG supplies from a variety of sources keeping markets well supplied and prices relatively low. This large projected increase in NG demand and production makes the experts to consider that a golden age of gas is coming [3].

With this new scenario, the existing 12.7 million compressed natural gas (CNG) vehicles on the world's roads will easily reach the predicted 50 million vehicles by 2020 [4]. In other words, NG is a strong candidate for replacing fuel oil as fuel for transportation. As combustible, NG produces significantly lower harmful emissions of nitrogen oxides, particulate matter, and greenhouse gas than oil-based gasoline or diesel. It does not produce VOC and sulfur oxides emissions, a common constituent of gasoline exhaust, but its use makes CO_2 emissions to rise. To avoid those CO_2 emissions, the best vehicular fuel is hydrogen (H_2), since the only emission product is water. However, hydrogen is not found free in the nature, but it must be produced from other natural sources, mainly from the steam reforming of NG. This fact increases the hydrogen fuel costs, which, from an economically point of view, makes difficult the transition to hydrogen vehicular fuel.

The main drawback of the use of NG (mainly methane) or H_2 as a substitute of fuel oil in the vehicles is their storage in the vehicular fuel tank. These gases are in a supercritical state at room temperature, so at these STP conditions, the energy density (defined as the heat of combustion per unit volume) is only 0.038 MJ/L (0.11% of that of gasoline) for NG and 0.0108 MJ/L

(0.03% of that gasoline) for H_2, so the mileage per unit volume of the fuel tank is very low. Hence, the suitability of NG and H_2 for vehicular application will depend on the ability to store an adequate amount in the onboard fuel tank. Commercially available cars, with a fuel tank of about 50 L, can cover a range of around 500 km. The Department of Energy of the United States (DOE) has established different targets for onboard NG and H_2 storage systems, including the minimum '"gravimetric" and "volumetric" capacities and the reversibility of the charging/discharging processes. Thus, to cover the same range than commercially available cars, the DOE storage target for methane has been defined as 150 volumes of stored methane at standard conditions (298 K and 0.1 MPa) per unit volume of vessel (i.e., 150 V/V) at 3.5 MPa. For hydrogen, the target is set at 6 wt% of stored hydrogen and a volumetric density of 0.045 kg H_2/L, to be achieved by 2010, with more ambitious targets of 9 wt% and volumetric density of 0.081 kg H_2/L set for 2015. These values are referred to the whole system, including the storage medium, the vessel, the refueling infrastructure, any regulators, electronic controllers, sensors, and so on (for more information, see www.eere.energy.gov/hydrogenandfuelcells). The more recently calculated values for "light-duty vehicles" [5] reduce the targets down to 4.5 wt% of H_2 and 0.028 kg H_2/L. In Europe, the targets are less restrictive and in the European Hydrogen and Fuel Cell Strategic Research Agenda and Deployment Strategy, Barret [6] points out an energy density value of 1.1 kWh L^{-1}, which is equivalent to a volumetric hydrogen storage capacity of about 0.033 kg H_2/L.

Several methods have been considered in order to increase the energy density of NG and H_2 and facilitate its use as a road vehicle fuel [7, 8]: (a) as a highly compressed gas (HCG), (b) as a liquid at very low temperature, and (c) as adsorbed gas in porous solids. Additionally, in the case of H_2, there is also another option: (d) forming metallic hydrides. In the case of NG, systems (a) and (b) are relatively easy to develop, especially the former, which is, as already commented, extensively used [4]. However, the use of these systems has some disadvantages. The storage tanks must be pressurized vessels (up to 20.7 MPa) and are thus constrained in their geometry (they are typically cylindrical), and are also rather heavy (<1 kg/L for steel tanks). Moreover, the compression itself is the most dangerous and complicated part of CNG systems. In the case of H_2, its storage at ambient temperature needs much higher pressures than that used for NG (20 to 60 MPa). Hence, in that case a costly multistage compression is required.

System (b), using cryogenic temperatures, can be applied to both supercritical gases (critical temperature of 191 K for methane and 32 K for H_2). Hence, such system involves the use of specialized and complicated container design (more in the case of H_2 than NG) as well as refueling procedures, which make more difficult their use in vehicles. More information about this type of systems can be found in a technical assessment of Cryo-Compressed Hydrogen Storage Tank Systems for Automotive Applications [9].

In the case of H_2, system (d) (use of metal hydrides) can also be applied. During absorption in metal hydrides, the hydrogen molecule is dissociated, its atoms diffuse into a metal host and form chemical bonds with it, resulting in the metal hydride phase. High volumetric capacities can be reached with metal hydrides, but, as it is based on an exothermal reaction, heat management is a drawback for this technique: A considerable amount of heat is generated upon hydrogenation, which has to be removed by an active cooling device. Thus, a number of secondary devices are needed, which again occupy precious space and lead to low storage capacities when the complete tank design is taken into account.

Some of above problems may be overcome with system (c). NG and H_2 can be stored as an adsorbed phase in porous materials filling the storage vessel. The search for a suitable porous material, in terms of further improving the storage volumetric energy density and lowering the adsorbent cost to the end use, is currently an active area of research. To obtain high storage capacity of NG and H_2, a variety of inorganic solids, such as zeolites, organic–inorganic hybrid porous complexes (MOF), and carbonaceous materials, such as superactivated carbons, activated carbon fibers, carbon monoliths, single-walled carbon nanotubes, and carbon nanohorns are being investigated [8–15]. They predominantly consist on microporous solids (pores of width around 0.8–2 nm). It has been shown [11] that zeolites have relatively high packing densities (compared with activated carbons) but have lower micropore volumes. In addition, zeolites are extremely hydrophilic and can lose their adsorption capacity for NG with time due to preferential moisture adsorption. Regarding MOF, although these porous solids do show a high adsorption capacity, their low tap density makes them unsuitable for these applications, since high container volumes would be required [16]. Additionally, high-pressure packing would increase the H_2 uptake, but for MOF the maximum pressure that can be used is restricted because its structure can collapse above 415 kg/cm^2 (for MOF-5) [16, 17]. Porous carbonaceous materials are characterized by their high adsorption capacity, good packing density, hydrophobic nature, and high adsorption/desorption rate [14, 15]. Moreover, both experimental [18] and theoretical [19] studies have shown that NG and hydrogen storage capacity is better in slit pores (typical pore shape in activated carbons) than in cylindrical ones (carbon nanotubes). Thus, carbon materials are very promising materials for NG and hydrogen storage [20].

In the case of hydrogen storage, carbon materials created great expectation largely due to the promising results reported by Dillon et al. [13]. It provoked the publication of an important number of studies using new carbon materials, such as single- and multi-wall carbon nanotubes and graphite carbon nanofibers prepared by chemical vapor deposition (CVD), arc discharge, laser vaporization, etc. [8, 10, 13, 15, 21–26]. Unfortunately, most of the outstanding values are not reproducible. As it has been analyzed in detail in the literature [8], the high result dispersion and low reproducibility of experimental measurements in carbon nanotubes are mostly consequence of experimental errors, such as non-purified samples, use of low-purity hydrogen, sonication bar material, and so on. On the other hand, for activated carbons and activated carbon fibers, hydrogen storage capacities of activated carbons measured by different authors are quite concordant, and data are less dispersed than in carbon nanotubes [15].

This chapter provides a summary of the results obtained by our research group over the past 15 years in the field of methane and hydrogen storage in carbon materials. The advantage of using our own results is that much better comparisons can be made and the main objective of the chapter can be reached more readily: to analyze the effect of the porosity (including pore volumes and pore size distributions) on methane and hydrogen storage capacities on a gravimetric basis. Moreover, the chapter points out the importance of the packing density of the materials in order to obtain high methane and hydrogen storage capacities on a volumetric basis, presenting results corresponding to activated carbon monoliths.

3.2 Adsorption of Methane on Activated Carbon

In this section, the performance in CH_4 (main component of NG) storage of several types of carbon materials is presented. Three morphologies of activated carbon materials prepared in our laboratory are included: (i) activated carbon fibers (ACFs) [27, 28], (ii) powdered superactivated carbons (PSACs) [29, 30], and (iii) cylindrical activated carbon monoliths (ACMs) [31]. The main objectives of this section are (a) to analyze the parameters of the activated carbons, which are important for methane adsorption, in order to optimize the material for methane storage applications and (b) to analyze, from a fundamental point of view, CH_4 adsorption under supercritical conditions (298 K and up to 4 MPa) on microporous carbon.

3.2.1 Optimization of Activated Carbons for Methane Storage Applications

The use of adsorbent materials in a storage vessel for storing methane, at relatively low pressures (3.5–4 MPa) and at room temperature, is a possibility for making NG vehicles competitive with other types of vehicles. As it has been previously said, in this application maximum methane storage density becomes the ultimate requirement, in order to store and deliver the maximum volume of gas per volume of storage vessel (V/V). Depending on the adsorbent's characteristics, the volumetric adsorption capacity and delivery will be different so does the volumetric energy density.

Enhancement of gas storage capacity through adsorption in carbon micropores occurs when the overall storage density is increased above that of the normal gas density at a given pressure. For example, Fig. 3.1 presents the methane adsorption capacity up to 4 MPa of a given adsorbent filling a tank (ANG) and of an empty tank pressurized at the same pressure. It can be observed that the methane adsorption capacity of the adsorbent filled tank is much higher than the empty tank. That happens when the adsorbed phase has a greater density than the gas phase in equilibrium with it. Consequently, there is an enhancement in a storage system of fixed volume because a greater amount of gas is adsorbed compared with the volume of gas displaced by the adsorbent volume. Following that, it can be said that in general, the higher the micropore volume of the adsorbent, the higher the methane adsorption capacity. This point can be analyzed in detail from the data compiled in Table 3.1. This table contains the porous texture characterization results (total micropore volume obtained from N_2 adsorption at 77 K ($V_{DR}(N_2)$), and narrow micropore volume (size smaller than around 0.7 nm) obtained from CO_2 adsorption at 273 K ($V_{DR}(CO_2)$) [30] and the gravimetric methane uptake (mg CH_4/g sample) for a wide variety of carbon materials: (i) powdered superactivated carbons (PSACs) prepared in our laboratory (series KUA), (ii) commercial powdered activated carbon (Maxsorb A), (iii) activated carbon fibers obtained from a carbon fiber prepared in our laboratory (series LF), and (iv) activated carbon fibers obtained from a commercial carbon fiber (series CF). It can be observed that the gravimetric methane uptake follows a good correlation with the total micropore volume.

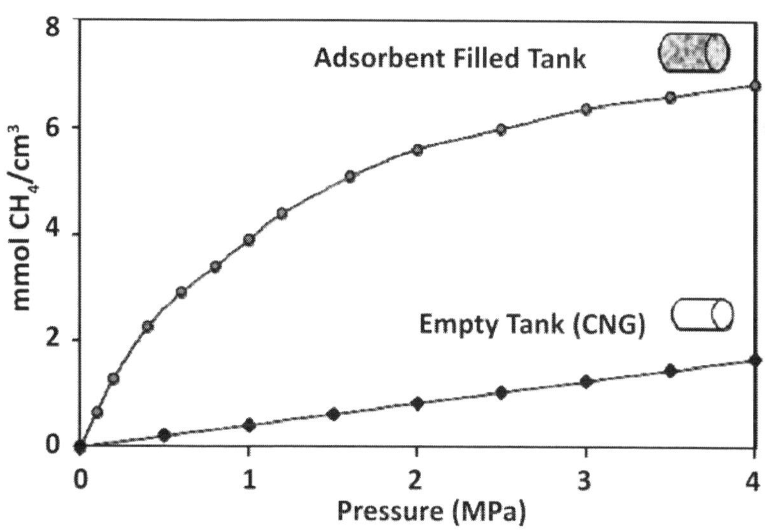

Figure 3.1 Comparison of the methane adsorption capacity of an adsorbent filled tank (ANG) and an empty tank pressurized up to 4 MPa at 298 K.

Table 3.1 Porous texture characterization results, packing density, gravimetric and volumetric methane uptake and delivery values for the PSACs prepared in our laboratory (series KUA), for a commercial powdered activated carbon (Maxsorb A), for the activated carbon fibers obtained from a carbon fiber prepared in our laboratory (series LF), and from a commercial carbon fiber (series CF)

Sample	BET (m^2/g)	$V_{DR}N_2$ (cm^3/g)	$V_{DR}CO_2$ (cm^3/g)	CH_4 uptake (mg/g)	Packing density (g/cm^3)	CH_4 uptake (V/V)	CH_4 delivered (V/V)
KUA11701	726	0.33	0.37	74	0.72	82	61
KUA21701	1305	0.58	0.54	119	0.69	127	99
KUA21751	2021	0.83	0.80	149	0.62	143	124
KUA21801	2085	0.96	0.81	164	0.63	160	126
KUA41651	2523	1.16	0.66	166	0.59	153	134
KUA41701	3290	1.45	0.81	191	0.56	166	142
KUA41751	3380	1.50	0.74	201	0.47	146	128
Maxsorb-A	3100	1.38	0.72	197	0.50	152	138
LFC30	930	0.41	0.65	92	0.90	126	99
LFC54	1930	0.85	0.75	136	0.76	155	135
LFC73	2400	0.98	0.70	154	0.70	163	143
LFS25	1050	0.46	0.48	64	0.98	102	72
LFS50	1560	0.65	0.48	102	0.80	128	108
CFC36	1079	0.53	0.46	84	1.06	136	107
CFC74	2862	0.92	0.62	148	0.74	166	150
CFS22	883	0.39	0.37	68	1.18	122	94
CFS64	1874	0.70	0.50	99	0.76	115	92

However, from an application point of view, in addition to the micropore volume, another crucial parameter, which has to be taken into account, is the packing density of adsorbent. This is so because in its real application, methane storage would be carried out in an activated carbon filled vessel. Then, obviously, the higher the amount of activated carbon in the vessel (which is related with the packing density), the higher the volume of methane stored. For example, in Fig. 3.2, these values of packing density of ACFs and PSACs have been plotted versus the micropore volume. Packing density of the samples was determined by pressing at 550 kg/cm^2 a given amount of activated carbon on a dry basis

(1 g approximately) in a mould. The measurements were repeated several times. The densities obtained have an error smaller than 5%. From these results, it can be said that, in general, the higher the porosity development of the materials, the lower the packing density, following a linear trend. It should be remarked that this type of porous materials, activated carbons, can be pressed up to 550 kg/cm^2, as it was done to determine the packing density, and even up to higher pressure, without changing their porous structure.

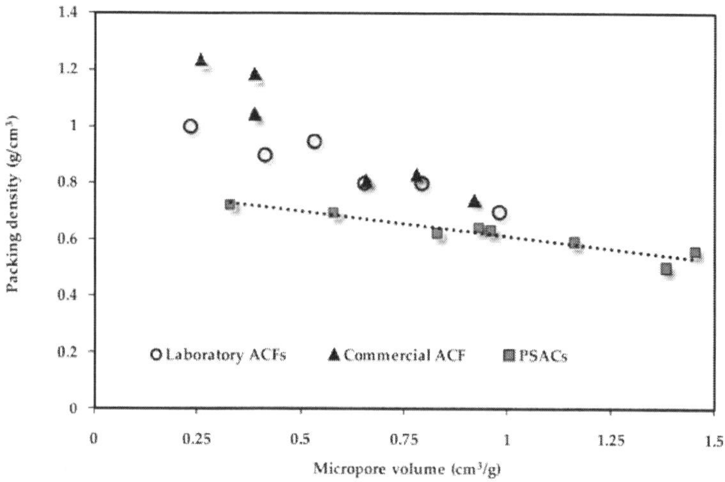

Figure 3.2 Packing density values corresponding to both ACFs and PSACs versus the micropore volume.

It can be seen in Fig. 3.2 that as expected, in both type of materials, the higher the micropore volume, the lower the packing density. It can be observed that the ACFs present higher packing density than the PSACs, especially for samples with micropore volume < 1 cm^3/g. This advantage is due, not only to the porous texture, but also to the morphology and flexibility of the fibers. This advantage of the ACFs compared to the PSACs is smaller as the percentage of activation increases. Using the measured packing densities, the volumetric methane adsorption uptake (V/V) in all the carbon materials was calculated from the gravimetric methane adsorption uptake and compiled in Table 3.1 for the series of ACFs and PSACs. It can be seen that even though one of the samples with the highest methane adsorption capacity on a gravimetric basis was the commercial activated carbon (Maxsorb-A), this sample does not have the highest volumetric adsorption capacity. The high packing density of the ACFs makes these materials to present higher volumetric uptake than PSACs, with similar gravimetric methane uptake.

From a practical point of view, the amount of methane that can be delivered by the activated carbon is more important than the total volumetric uptake capacity. If the way of recovering the adsorbed methane is just by decreasing the gas pressure from 4 to 0.1 MPa, the amount of methane delivered can be obtained as the difference between the amount of methane adsorbed at 4 MPa and the amount retained at 0.1 MPa. This parameter is defined as the volume of gas delivered per volume of the storage container is shown in Table 3.1 for the series of ACFs and PSACs. It can be observed that the methane delivery values are 9–29% lower than the volumetric methane adsorption capacity. Thus, sample KUA21801 has one of the highest volumetric methane uptake (160 V/V), but its delivery is not very high (126 V/V). The relatively high retention of methane at atmospheric pressure of this sample can be explained considering that it presents a large amount of narrow micropores in its MPSD. This can be deduced from the comparison between the micropore volumes given by $V_{DR}(N_2)$ and $V_{DR}(CO_2)$ [32]. A small difference between both micropore volumes ($V_{DR}(N_2)$ - $V_{DR}(CO_2)$) indicates that most of the micropores of the sample are narrow with a pore size close to 0.7 nm, while a higher difference between both values indicates a wider MPSD and a higher average micropore size. The detailed analysis of these results seems to indicate that samples with too narrow pore size are not suitable for methane storage applications because they retain a lot of methane at atmospheric pressure [27].

Thus, as a summary, it can be concluded that, in order to get the maximum CH_4 delivery, we need samples with high micropore volume, a suitable MPSD, and a high packing density. However, it has been seen (Fig. 3.2) that some of these parameters (micropore volume and packing density) are inversely related, so that the best balance between these two parameters should be found.

Owing to the enormous importance of the packing density of the activated carbons for this application, the useless voids (meso, macro, and interparticle space), where methane adsorption does not take place in an important amount, must be minimized, while maintaining a high micropore volume. In the case of mainly microporous PSACs, which do not present macroporosity, as is the case of the activated carbon materials presented previously, this void consists of any space between the particles. By producing the activated carbon as a monolith (ACMs), it is possible to reduce that interparticle voids, maximizing the bulk density [33, 34]. These monoliths consist of cylindrically shaped pieces, which pack uniformly, lessening wasted space in the storage vessel. Another advantage of producing carbons in an efficient space-filling form is that they are also strong and resist attrition. The monoliths can be produced using a binder, which helps to keep the carbon particles in a compressed state. Conventional preparation methods [34, 35] consist of mixing the activated carbon with a binder, compression and molding using a hydraulic press and, finally, pyrolysis to improve the binder properties and decrease the weight of binder in the monolith. For methane storage application, the binder must possess certain properties: (i) It must produce monoliths with good mechanical properties using the lowest binder/activated carbon ratio to achieve the maximum amount of activated carbon, which is the adsorbent material and (ii) its pyrolysis must not produce pore blocking of the activated carbon. This last characteristic of the binder is very important, because the current focus is to increase the packing density, while retaining a high micropore volume, which is responsible for the methane adsorption.

The effect of the type and amount of binder on the micropore volume, methane adsorption capacity, piece density and delivery was analyzed in detail in a previous work [31]. Different binders were used to prepare ACMs: a humic acid-derived sodium salt (HAS) from Acros Organics, polyvinyl alcohol (PVA), a phenolic resin (PR) (Georgia Pacific 5506), a propriety binder from Waterlink Sutcliffe Carbons (WSC), Teflon (TF) from Du Pont, and a cellulose-based binder (ADH). For the preparation of these ACMs, the best PSAC prepared in our laboratory for methane storage applications (delivery 145 V/V) was selected. The study of the agglomeration of this PSAC has shown that most of the binders studied produce monoliths with good mechanical properties. However, the adsorption capacities of the ACMs are reduced with respect to the starting activated carbon. This reduction is due to the reduction in the amount of activated carbon in the monolith and to a partial blocking of the porosity, which depends on the kind of binder used for the preparation of the monoliths. Among all the binders studied, the one that produces monoliths with the best equilibrium between adsorption capacity and piece density, and consequently with the best uptake (140 V/V) and delivery (126 V/V), is the binder WSC.

3.2.2 Correlation of Methane Adsorption with the Micropore Texture

The existence of a reliable correlation between methane adsorption and the porous texture of activated carbons is of interest as it would enable to check easily the suitability of a given material if high-pressure equipment is not available. In the literature, various correlations have been proposed between methane adsorption at 298 K and porosity parameters obtained from N_2 adsorption at 77 K [27, 31, 33]. In an earlier work, it was shown that the linear trend obtained in the plot of CH_4 adsorption capacity versus parameters such as BET surface area and micropore volume obtained from N_2 adsorption at 77 K fails for those ACFs that have narrow microporosity [27]. In that work [27], it was concluded that the best correlation between methane storage capacity of an ACF and porosity is found with the total micropore volume.

In Fig. 3.3a, CH_4 uptake is plotted versus the total micropore volume for the series of samples PSACs and ACFs (calculated from N_2 DR plots up to relative pressures of 0.14) [36]. As it can be seen, most of the samples follow a linear trend (up to 1 cm^3/g) passing through the origin. In the case of PSACs with micropore volumes higher than 1 cm^3/g CH_4 uptake deviation occurs. This fact should be related to the existence of supermicroporosity and, consequently, to a lower CH_4 density in such porosity compared to narrow micropores. Applying the DR equation to the N_2 data only in the adsorption potential range where the characteristic curves for CH_4, N_2 and CO_2 fit (>100 $(kJ/mol)^2$), corresponding to relative pressures of N_2 below 0.07) [36], we can calculate different micropore volumes, which are plotted in Fig. 3.3b. A much better correlation between CH_4 uptake and V_{micro} can be observed in this plot. The new V_{micro} obtained differs from the previous ones (Fig. 3.3a) only for those samples that have wide MPSDs. In these samples, all their V_{micro} values (estimated using N_2 DR plots up to relative pressures of 0.14) are not useful for CH_4 adsorption at 4 MPa and room temperature since their wide pore sizes.

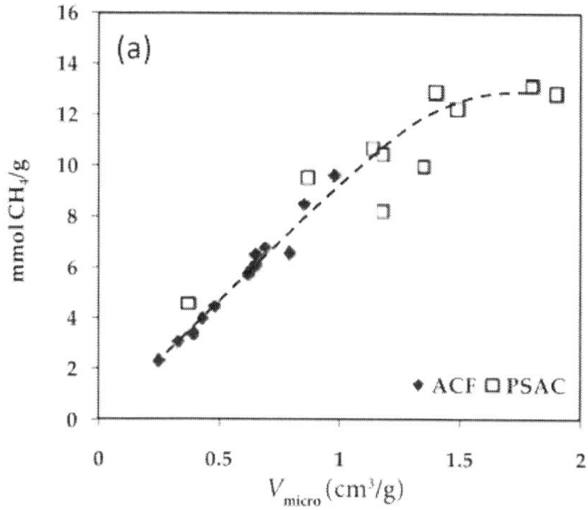

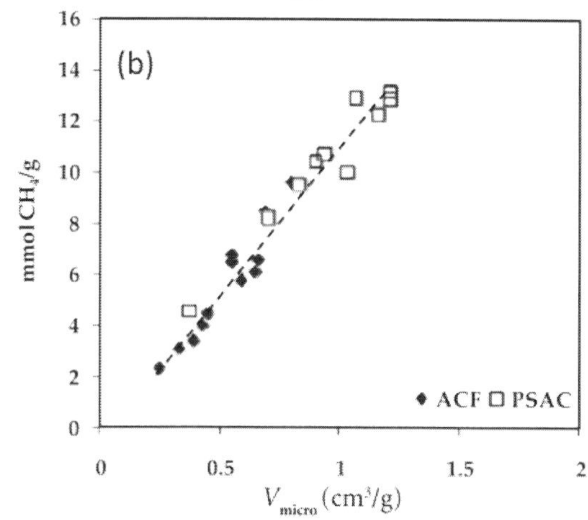

Figure 3.3 CH_4 uptake at 4 MPa versus: (a) total micropore volume, (b) micropore volume obtained up to $P/P_s = 0.07$ of N_2 adsorption data.

Frequently, parameters affecting high-pressure methane adsorption have been analyzed in the literature [33, 34, 37] using the BET equation. For example, Clarkson et al. [37] tested, in addition to the adsorption potential (DR), the validity of mono/multilayer concepts (Langmuir and BET theories). Both the DR and BET equations fit very well the experimental data. To deepen into the suitability of the use of the BET equation at high-pressure methane adsorption, Fig. 3.4a plots the monolayer capacities (n_m) obtained by application of the BET equation to the CH_4 isotherms

[36] together with the experimental amount of methane uptake at 4 MPa. Interestingly, in general, the calculated n_m values are quite similar to the experimental methane uptake values. Similarly, Fig. 3.4b presents a plot of monolayer capacities (n_m) (from CH_4 adsorption data) versus the N_2 BET surface area. As it can be seen, most of the samples follow a linear trend. The very good relationships between n_m-BET and the total CH_4 uptake at 4 MPa, as well as CH_4 BET, confirm that BET equation is a good tool to measure the adsorption capacity of microporous solids (i.e., N_2 at 77 K or CH_4 at 298 K and 4 MPa).

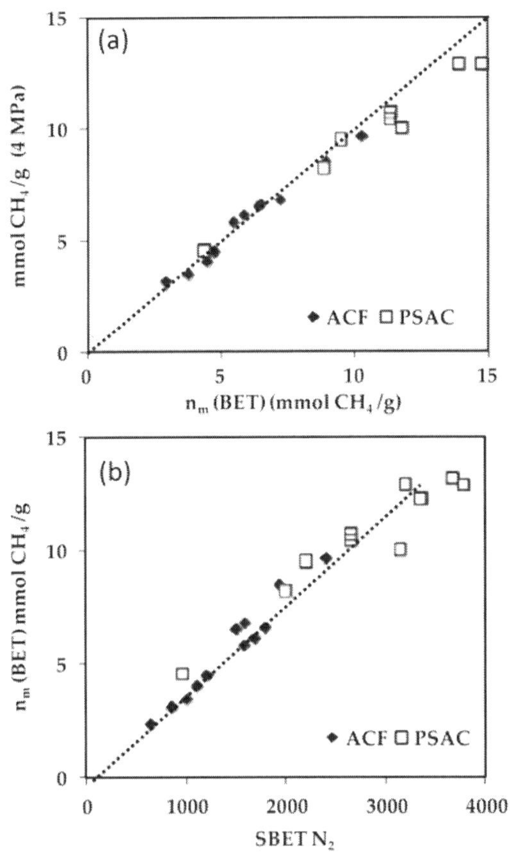

Figure 3.4 Monolayer capacities (n_m) obtained by applying the BET equation to CH_4 isotherms versus: (a) the experimental amount of methane uptake at 4 MPa; (b) Surface area of activated carbons.

3.3 Adsorption of Hydrogen on Activated Carbons

In the same way that it has been done for methane storage, in order to find the best microporous carbon and the adsorption conditions that provide the highest hydrogen storage, it is important to find correlations between the hydrogen adsorption data at different conditions (temperatures and pressures) and the properties of the carbon materials (specific surface area, pore size distribution, structure and packing density). Thus, in this section, experimental results corresponding to hydrogen adsorption at high pressures (up to 20 MPa) and different temperatures (298 K and 77 K) for a wide variety of well-characterized carbon materials is presented. These carbon materials include activated carbons, activated carbon fibers, ACMs, carbon nanotubes, and carbon nanofibers. Moreover, the importance of the packing density of the materials for both, comparison between samples and also from an application point of view (to know their real hydrogen storage capacity) is pointed out.

3.3.1 Correlation of Hydrogen Adsorption with the Micropore Texture

Different authors, analyzing hydrogen adsorption at 77 K, have reported that for porous carbons the total amount of hydrogen uptake per gram of adsorbent is correlated with the porosity development of the material. On the contrary, for hydrogen adsorption capacity at room temperature, the pore size distribution of the sample positively contributes, as shown in our previous work [38] and also by Ströbel et al. [39]. A closer look into the literature shows important discrepancies on the interpretation of the experimental data. Thus, some researchers have established that hydrogen adsorption capacity of porous carbons is directly related to the BET surface area and total micropore volume [23, 40], while others have not found a so clear relationship between those parameters [39, 41]. On the other hand, other researchers have found a linear trend between hydrogen adsorption and the narrow micropore volume (determined by CO_2 adsorption at 273 K) [42, 43]. In a previous paper [15] we provided an interpretation of the different correlations found, using the fundamentals of adsorption at supercritical conditions and a wide variety of carbon materials and adsorption conditions. To summarize, in the present review, the results corresponding to selected samples are presented, which include (i) chemically activated multi-wall carbon nanotubes; (ii) commercially available vapor grown carbon nanofibers (NF); (iii) sample NF after chemical activation with KOH; (iv) a series of activated carbons prepared from a Spanish anthracite by chemical activation with KOH (samples named KUA1, KUA5, and KUA6); (v) commercial pitch-based carbon fiber from Kureha Company (sample CF); (vi) commercially available activated carbon AX-21 from Anderson Carbon Co.; (vii) commercial activated carbon fibers from Osaka Gas Co. (sample A20). The detailed porosity characterization results of these samples can be found elsewhere [15].

Figure 3.5a shows the amount of hydrogen adsorbed (i.e., adsorption excess) at 20 MPa and room temperature (298 K), versus the total micropore volume ($V_{\text{micro}}(N_2)$) for the selected carbon materials. A general trend, where hydrogen adsorption capacity increases with total micropore volume can be observed. However, some deviations to this general trend exist. It can be seen that, at high micropore volume values, the sample with the highest porosity development (sample KUA6) does not present the highest hydrogen adsorption capacity. The highest hydrogen adsorption capacity value obtained in the present study is 1.2 wt% and corresponds to the sample KUA5. This sample has a very high micropore volume and also a quite narrow MPSD, as it can be deduced for the relatively low value of the difference ($V_{\text{micro}}(N_2) - V_{\text{micro}}(CO_2) = 0.38$ cm^3/g) (see reference [15]). These results point out the importance of developing materials having both high micropore volume and narrow MPSD for the adsorption of hydrogen at room temperature. This conclusion agrees with that obtained in a previous paper that showed that the optimum pore size for hydrogen adsorption on porous carbons is close to 0.6 nm [19]. Thus, plotting, for this series of samples that also include the vapor grown carbon nanofibers and the nanotubes (triangle symbols), the hydrogen adsorption capacity values versus the narrow micropore volume ($V_{\text{micro}}(CO_2)$) (Fig. 3.5b), a slight better linear trend is observed.

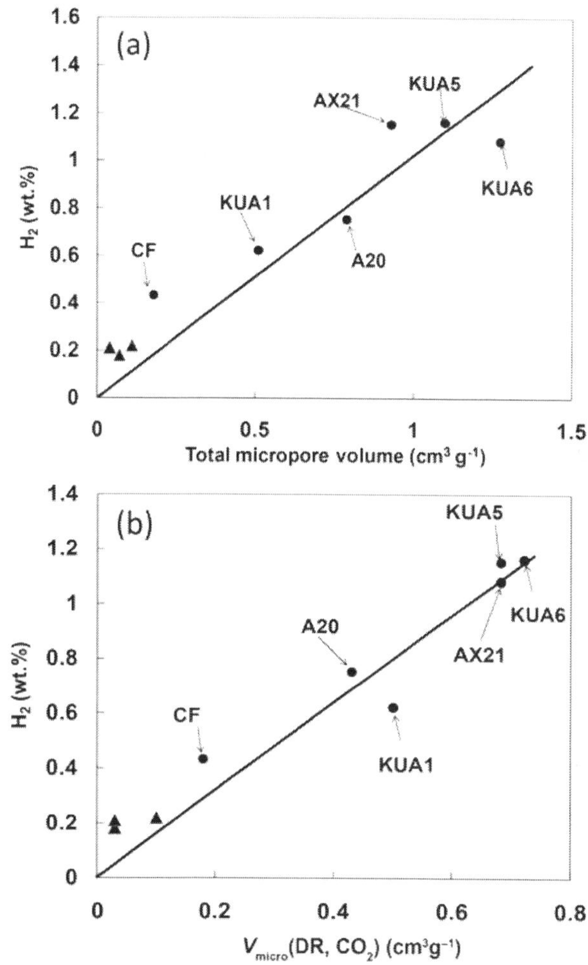

Figure 3.5 Amount of hydrogen adsorbed at 298 K and 20 MPa by activated carbons (circles) and carbon nanotubes and nanofibres (triangles) versus the total micropore volume (a) and the narrow micropore volume (V_{micro}(DR, CO_2)) (b).

In order to emphasize the importance of developing materials with a tailored porosity for hydrogen adsorption at room temperature and high pressure, the hydrogen adsorption isotherm for the sample KUA5 has been measured at 298 K and up to 50 MPa, using the same experimental system than the one used in a previous study [19]. Figure 3.6a plots the hydrogen adsorption

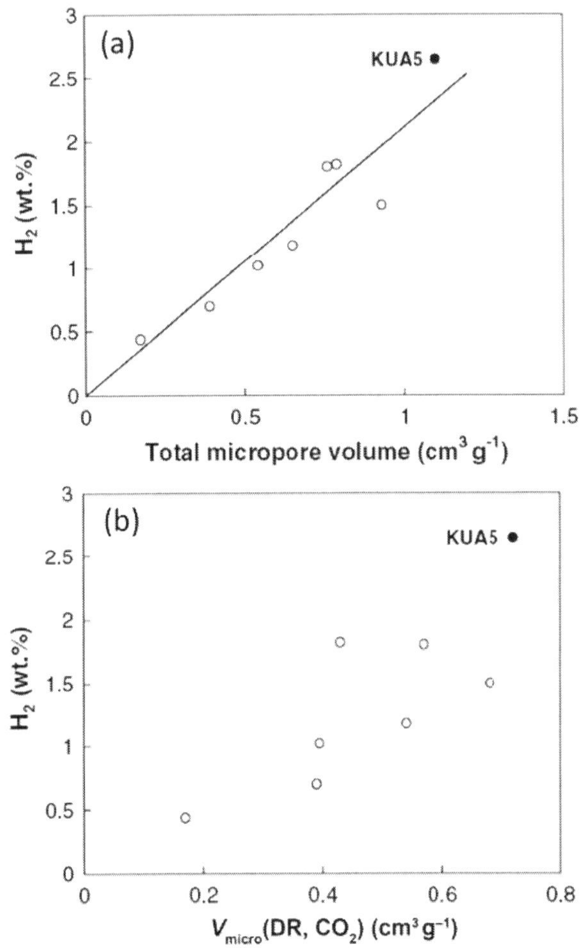

Figure 3.6 Amount of hydrogen adsorbed at 298 K and 50 MPa versus the total micropore volume: (a) and the narrow micropore volume (V_{micro}(DR, CO_2)); (b) (values found in the literature [16] (open symbols) and sample KUA5 (closed symbol)).

capacity at 50 MPa versus the total micropore volume ($V_{micro}(N_2)$) for data already published [19] and for sample KUA5. It can be seen that hydrogen adsorption capacity at 50 MPa also follows a linear relationship with the total micropore volume. The chemically activated carbon KUA5 presents a much higher adsorption capacity than the samples of the previous study, reaching a value of 2.7 wt%. Contrary to what happened in the case of the hydrogen adsorption capacity at 20 MPa, if hydrogen adsorption capacity at 50 MPa is plotted versus the narrow micropore volume (see Fig. 3.6b), a very high dispersion of the results is obtained. The fact that only a good correlation is found with total micropore volume can be under-stood, considering that at 50 MPa the contribution of the wider micropores to the total adsorption is important.

Figure 3.7 includes the hydrogen adsorption capacity at 77 K for the selected materials. It can be seen that this series of samples covers a wide range of hydrogen adsorption capacity reaching, for the best samples, values higher than 5 wt% hydrogen (expressed as $m_{H_2}/(m_C + m_{H_2})$) at relatively low pressures (3–4 MPa) (samples KUA5 and KUA6). Similar to what happened at room temperature, the lowest hydrogen uptake values correspond to the nanofibers and nanotubes samples, according to their lower porosity development, as shown in our previous work [15]. In the case of hydrogen adsorption at room temperature, a linear relationship is obtained between the amount of hydrogen adsorbed at 77 K and the porosity of the samples. As it can be observed in Fig. 3.7a, a good correlation is obtained between the amount of hydrogen adsorbed at both 0.1 MPa and 4 MPa and the total micropore volume. However, contrary to the results obtained at room temperature, in this case, the highest adsorption capacity corresponds to the KOH-activated carbon with the highest development of porosity (sample KUA6). That means that at 77 K, no effect of the MPSD on the hydrogen adsorption capacity has been observed, when the adsorption is done in this range of pressures. When the hydrogen adsorption capacity at 77 K is plotted versus the narrow micropore volume ($V_{micro}(CO_2)$) (Fig. 3.7b), it can be seen that the correlations are worse compared with those with the total micropore volume, although the best one is obtained when the amount of hydrogen adsorbed at low pressures (0.1 MPa) is used. As observed in Fig. 3.7, the results obtained for carbon nanofibers and carbon nanotubes (solid symbols) fit in the tendencies obtained for activated carbons, showing that hydrogen adsorption at these conditions only depends on the porosity of the samples.

From these results, it can be concluded that hydrogen adsorption at 77 K and at pressures up to 4 MPa requires porous solids with a very high micropore volume, such as chemically activated carbons. At these adsorption conditions, MPSD does not play an important role, contrary to what happens at room temperature.

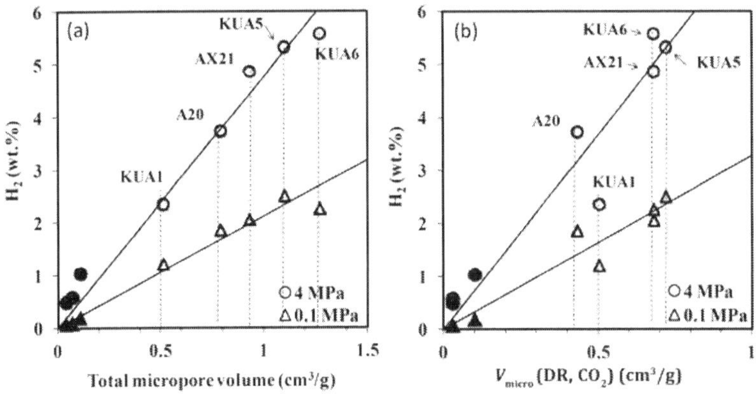

Figure 3.7 Amount of hydrogen adsorbed at 77 K and different pressures (0.1 and 4 MPa) by activated carbons (open symbols) and carbon nanotubes and nanofibres (closed symbols) versus the total micropore volume (a) and the narrow micropore volume [V_{micro} (DR, CO_2)] (b).

3.3.2 Hydrogen Adsorption in Volumetric Basis and Total Hydrogen Storage Capacity

Our previous study [44] pointed out the importance, from an application point of view, of reporting the hydrogen adsorption capacities on a volumetric basis (estimated using the packing density values). It was shown that when plotting the hydrogen adsorption capacity at 77 K on a volumetric basis versus the micropore volume, the maximum does not correspond to the sample with the highest micropore volume, as it was observed in the case of hydrogen adsorption on a gravimetric basis (Fig. 3.7), but corresponds to a sample with a micropore volume around 1 cm^3/g, which is a sample that still has a quite high packing density, as seen in Fig. 3.2. Thus, analogous to NG storage, in order to get a maximum hydrogen adsorption capacity on a volumetric basis (the real key point for its application in a vehicle), we should choose an activated carbon with a good balance between porosity development, and packing density. As mentioned in Section 3.2.1, from an application point of view, ACMs are much more interesting than powder samples. A systematic study about preparation of monoliths (typical monoliths dimensions: 1.6 cm diameter and approximately 1 cm height) carried out in our laboratory has allowed us to obtain ACM with high performance in hydrogen storage, as shown elsewhere [44]. It was shown how ACM can be prepared with similar porosity than powder activated carbon but with higher density. Thus, for example, powder activated carbon with a micropore volume of 0.93 cm^3/g has a packing density of around 0.54 g/cm^3. On the other hand, in the case of ACM, pieces of carbon with similar micropore volume (0.98 cm^3/g) can reach higher densities (around 0.6 g/cm^3). This increase of the density of the material gives a value of hydrogen adsorption on a volumetric basis for an ACM of 29.7 g H_2/L at 77 K and at 4 MPa. In addition to porous texture characterization and piece density estimation, the ACM was submitted to a complementary mechanical characterization, and it was concluded that they have very good mechanical properties and present a low pressure drop and a low thermal expansion.

The results presented so far both on gravimetric and volumetric basis correspond to the adsorption excess of hydrogen on different solids at different temperatures and pressures. However, a more important parameter from an application point of view, which is rarely found in the literature, is the total storage capacity. The storage capacity in a specified volume filled with a physisorption-based hydrogen carrier is the sum of the capacity due to adsorption on the solid surface and the volumetric capacity due to compression in the void space [45]. The void space per unit volume (V_s) was assessed using the packing density (ρ_p) and the skeleton density (ρ_s) of adsorbent ($V_s = 1 - (\rho_p/\rho_s)$). The packing density of the materials was measured as described in the previous section. The skeleton density of the activated carbons was determined by the helium expansion method. Figure 3.8 includes the total hydrogen storage capacity of the SACs (sample KUA5) at 298 K and at 77 K on the basis of a 1 L container. These figures also include the amount of hydrogen stored just by compression and the adsorption excess data, based on the weight of carbon in the 1 L container (the packing density of activated carbon KUA5 is 0.5 g/cm^3). It can be seen that storage of hydrogen at room temperature and 77 K achieves considerable enhancement due to both adsorption and compression. The total storage capacity of a 1 L container filled with carbon powder KUA5 is 16.7 g H_2 at 19.5 MPa and 298 K and 38.8 g H_2 at 4 MPa and 77 K. In the case of the best sample prepared obtained in our laboratory in terms of volumetric values (the ACM), the total storage capacity of a 1 L container filled with this material is 39.3 g H_2 at 4 MPa and

77 K. Interestingly, this value of volumetric hydrogen storage capacity is almost 1.2 times higher than the European target for on-board hydrogen storage systems (33 g H_2/L). Considering the good results obtained with these materials, our effort now focuses on optimizing monoliths preparation and properties.

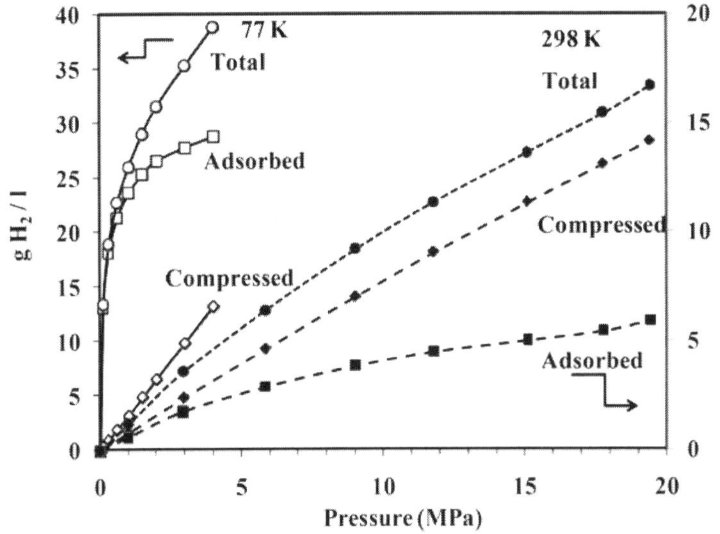

Figure 3.8 Total hydrogen storage capacity of activated carbon KUA5 on the basis of a 1 L container at 298 K and 77 K.

If the weight percentage of total H_2 stored on activated carbon KUA5 at 298 K is assessed from the hydrogen adsorption isotherms measured up to 50 MPa, the values obtained are 3.2 wt% and 6.8 wt% at 20 MPa and 50 MPa, respectively. Very high pressures (50 MPa) are required to approach to the DOE target at room temperature with a KOH-activated carbon. On the other hand, at 77 K, the samples KUA5 and KUA6 gave very interesting values of H_2 stored at a quite low pressure (4 MPa), 7.1 wt% and 8.0 wt%, respectively.

It can be observed that the preparation of a series of KOH-activated carbons with a wide range of properties has allowed us to select very good materials for different hydrogen adsorption conditions, remarking the importance of developing materials with suitable porosity for this application. All these results confirm that activated carbons are promising candidates for hydrogen storage for onboard vehicles.

3.4 Conclusions

Experimental methane and hydrogen adsorption results at different temperatures and pressures on a series of chemically activated carbons prepared in our laboratory with a wide range of porosities such as ACMs, activated carbon fibers, carbon nanotubes, and vapor grown carbon nanofibers have been presented. This chapter provides a useful interpretation of methane and hydrogen adsorption data according to the porosity of the materials and to the adsorption conditions. It has been observed that the gravimetric methane uptake follows a good correlation with the micropore volume obtained up to $P/P_s = 0.07$ of N_2 adsorption data. In the case of hydrogen, it has been concluded that at 298 K, the hydrogen adsorption capacity depends on both the micropore volume and the MPSD, whereas at 77 K, hydrogen adsorption depends on the surface area and the total micropore volume of the activated carbon.

In this chapter, it is also remarked that from an application point of view and also for purposes of comparison between different types of porous materials, it is important to measure and report the packing density of the materials in order to obtain the methane and the hydrogen adsorption capacities on a volumetric basis. The results point out the importance of developing materials that have both high micropore volume and high packing density, in order to get a high methane and hydrogen adsorption capacity on a volumetric basis. Methane delivery up to 150 V/V has been obtained for an activated carbon fiber prepared in our laboratory. The total hydrogen storage on the best powder activated carbon at 298 K is 16.7 g H_2/L and 37.2 g H_2/L at 20 MPa and 50 MPa, respectively and 38.8 g H_2/L at 77 K and 4 MPa.

Taking into account the advantages of monoliths for methane and hydrogen storage applications (easier to handle than powder, good mechanical properties, and high density), we have prepared ACMs that have both high micropore volume and high density. The optimization of the preparation conditions has allowed us to obtain an ACM with a methane uptake at 3.5 MPa of 140 V/V and a delivery of 126 V/V. Moreover, another ACM has been optimized for hydrogen storage. That material presents a hydrogen adsorption capacity of 29.7 g H_2/L at 77 K and 4 MPa. The total storage capacity of a 1 L container filled with that ACM has been estimated to be

39.3 g H_2 at 4 MPa and 77 K. This value of volumetric hydrogen storage capacity is almost 1.2 times higher than the European target for on-board hydrogen storage systems (33 g H_2/L). The results presented in this chapter confirm that activated carbons are promising candidates for NG and hydrogen storage for onboard vehicles.

Acknowledgments

The authors would like to thank the Spanish Ministerio de Ciencia e Innovación and PLAN E funds (Project CTQ2009-10813/PPQ) and Generalitat Valenciana and FEDER (PROMETEO/2009/047) for financial support.

References

1. Report DOE/EIA-0484 (2010).
2. (a) Sunggyu, L., Speight, J.G., and Loyalka, S.K. (2007). *Handbook of Alternative Fuel Technologies* (CRC Press, Boca Raton, FL); (b) Ngô, C., and Natowitz, J.B. (2009). *Our Energy Future: Resources, Alternatives and the Environment* (John Wiley & Sons, Inc., Hoboken, New Jersey, USA).
3. *Are We Entering A Golden Age of Gas?*, World Energy Outlook 2011—special report 2011, IEA, http://www.worldenergyoutlook.org/.
4. NGV Global Association Website Operational, http://www.ngvglobal.org/.
5. http://www1.eere.energy.gov/hydrogenandfuelcells/storage/pdfs/targets_onboard_hydro_storage.pdf.
6. Barrett, S. (2005), *Fuel Cells Bull.*, p. 1.
7. CooK, T.L., Komodromos, C., Quinn, D.F., and Rangan, S. (1999). *Carbon Materials for Advanced Technologies*, ed., Burchell, T.D., Chapter 9, "Adsorbent Storage for Natural Gas Vehicles" (Pergamon, New York) pp. 269–302.
8. Schlapbach, L., and Züttel, A. (2001). Hydrogen-storage materials for mobile applications, *Nature*, **414**, pp. 353–358.
9. http://www1.eere.energy.gov/hydrogenandfuelcells/pdfs/cyro_compressed_auto.pdf.
10. Berry, G.D., and Aceves, S.M. (1998). Onboard storage alternatives for hydrogen vehicles, *Energy Fuels*, **12**, pp. 49–55.

11. Menon, V.C., and Komarneni, S.J. (1998). Porous adsorbents for vehicular natural gas storage: a review, *Porous Mater.*, **5**, pp. 43–58.

12. Rosi, N.L., Eckert, J., Eddaoudi, M., Vodak, D.T., Kim, J., O'Keeffe, M., and Yaghi, O.M. (2003). Hydrogen storage in microporous metal-organic frameworks, *Science*, **300**, pp. 1127–1129.

13. Dillon, A.C., Jones, K.M., Bekkedahl, T.A., Kiang, C.H., Bethune, D.S., and Heben, M.J. (1997). Storage of hydrogen in single-walled carbon nanotubes, *Nature*, **386**, pp. 377–379.

14. Lozano-Castelló, D., Alcañiz-Monge, J., de la Casa-Lillo, M.A., Cazorla-Amorós, D., and Linares-Solano, A. (2002). Advances in the study of methane storage in porous carbonaceous materials, *Fuel*, **81**, pp. 1777-1803.

15. Jordá-Beneyto, M., Suárez-García, F., Lozano-Castelló, D., Cazorla-Amorós, D., and Linares-Solano, A. (2007). Hydrogen storage on chemically activated carbons and carbon nanomaterials at high pressures, *Carbon*, **45**, pp. 293–303.

16. Juan-Juan, J., Marco-Lozar, J.P., Suárez-García, F., Cazorla-Amorós, D., and Linares-Solano, A. (2010). A comparison of hydrogen storage in activated carbons and a metal-organic framework (MOF-5), *Carbon*, **48**, pp. 2906–2909.

17. Bhatia, S.K., and Myers, A.L. (2006). Optimum conditions for adsorptive storage. *Langmuir*, **22**, pp. 1688–1700.

18. Zhou, L., Zhou, Y.P., and Sun, Y. (2004). A comparative study of hydrogen adsorption on superactivated carbon versus carbon nanotubes, *Int. J. Hydrogen Energy*, **29**, pp. 475–479.

19. Rzepka, M., Lamp, P., and de la Casa-Lillo, M. (1998). Physisorption of hydrogen on microporous carbon and carbon nanotubes, *J. Phys. Chem. B*, **102**, pp. 10894–10898.

20. Linares-Solano, A., Cazorla-Amorós, D., Marco-Lozar, J.P., and Suárez-García, F. (2011). *Coordination Polymers and Metal Organic Frameworks*. Chapter 6, "High pressure gas storage on porous solids; a comparative study of MOFs and Activated Carbons" (Nova Science Publishers, Inc.).

21. Dillon, A.C., and Heben, M.J. (2001). Hydrogen storage using carbon adsorbents: past, present and future, *Appl. Phys. A-Mater. Sci. Process.*, **72**, pp. 133–142.

22. Liu, C., Fan, Y.Y., Liu, M., Cong, H.T., Cheng, H.M., and Dresselhaus, M.S. (1999). Hydrogen storage in single-walled carbon nanotubes at room temperature, *Science*, **286**, pp. 1127–1129.

23. Takagi, H., Hatori, H., Soneda, Y., Yoshizawa, N., and Yamada, Y. (2004). Adsorptive hydrogen storage in carbon and porous materials, *Mater. Sci. Eng. B-Solid State Mater. Adv. Technol.*, **108**, pp. 143–147.

24. Chambers, A., Park, C., Baker, R.T.K., and Rodriguez, N.M. (1998). Hydrogen storage in graphite nanofibers, *J. Phys. Chem. B*, **102**, pp. 4253–4256.

25. Chen, P., Wu, X., Lin, J., and Tan, K.L. (1999). High H_2 uptake by alkali-doped carbon nanotubes under ambient pressure and moderate temperatures, *Science*, **285**, pp. 91–93.

26. Hirscher, M., Becher, M., Haluska, M., Quintel, A., Skakalova, V., Choi, Y.M., Dettlaff-Weglikowska, U., Roth, S., Stepanek, I., Bernier, P., Leonhardt, A., and Fink, J. (2002). Hydrogen storage in carbon nanostructures, *J. Alloys Compounds*, **330–332**, pp. 654–658.

27. Alcañiz-Monge, J., de la Casa-Lillo, M.A., Cazorla-Amorós, D., and Linares-Solano, A. (1997). Methane storage in activated carbon fibres, *Carbon*, **35**, pp. 291–297.

28. Alcañiz-Monge, J., Cazorla-Amorós, D., Linares-Solano, A., Oya, A., Sakamoto, A., and Hoshi, K. (1997). Preparation of general purpose carbon fibers from coal tar pitches with low softening point, *Carbon*, **35**, pp. 1079–1089.

29. Lozano-Castelló, D., Cazorla-Amorós, D., Linares Solano, A., and Quinn, D.F. (2002). Influence of pore size distribution on methane storage at relatively low pressure: preparation of activated carbon with optimum pore size, *Carbon*, **40**, pp. 989–1002.

30. Lozano-Castelló, D., Lillo-Ródenas, M.A., Cazorla-Amorós, D., and Linares-Solano, A. (2001). Preparation of activated carbons from Spanish anthracite—I. activation by KOH, *Carbon*, **39**, pp. 741–749.

31. Lozano-Castelló, D., Cazorla-Amorós, D., Linares-Solano, A., and Quinn, D.F. (2002). Activated carbon monoliths for methane storage: influence of binder, *Carbon*, **40**, pp. 2817–2825.

32. Rodríguez-Reinoso, F., and Linares-Solano, A. (1988). *Chemistry and Physics of Carbon* **21**, ed., Thrower, P.A. "Microporous structure of activated carbons as revealed by adsorption methods" (Marcel Dekker, New York), pp. 1–146.

33. Parkyns, N.D., and Quinn, D.F. (1995). *Porosity in Carbons*, ed., Patrick, J.W., Chapter 11, "Natural gas adsorbed on carbon," (Edward Arnold, London) pp. 293–325.

34. Quinn, D.F., and MacDonald, J.A. (1992). Natural gas storage, *Carbon*, **30**, pp. 1097–1103.

35. Chen, X.S., McEnaney, B., Mays, T.J., Alcañiz-Monge, J., Cazorla-Amorós, D., and Linares-Solano, A. (1997). Theoretical and experimental studies of methane adsorption on microporous carbons, *Carbon*, **35**, pp. 1251–1258.

36. Alcañiz-Monge, J., Lozano-Castelló, D., Cazorla-Amorós, D., and Linares Solano, A. (2009). Fundamentals of methane adsorption in microporous carbons, *Microporous Mesoporous Mater.*, **124**, pp. 110–116.

37. Clarkson, C.R., Bustin, R.M., and Levy, J.H. (1997). Application of the mono/multilayer and adsorption potential theories to coal methane adsorption isotherms at elevated temperature and pressure, *Carbon*, **35**, pp. 1689–1705.

38. de la Casa-Lillo, M., Lamari-Darkrim, F., Cazorla-Amorós, D., and Linares-Solano, A. (2002). Hydrogen storage in activated carbons and activated carbon fibers, *J. Phys. Chemis. B*, **106**, pp. 10930–10934.

39. Ströbel, R., Jorissen, L., Schliermann, T., Trapp, V., Schutz, W., Bohmhammel, K., Wolf, G., and Garche, J. (1999). Hydrogen adsorption on carbon materials, *J. Power Sources*, **84**, pp. 221–224.

40. Panella, B., Hirscher, M., and Roth, S. (2005). Hydrogen adsorption in different carbon nanostructures, *Carbon*, **43**, pp. 2209–2214.

41. Nijkamp, M.G., Raaymakers, J.E.M.J., van Dillen, A.J., and de Jong, K.P. (2001). Hydrogen storage using physisorption—materials demands, *Appl. Phys. A-Mater. Sci. Process.*, **72**, pp. 619–623.

42. Texier-Mandoki, N., Dentzer, J., Piquero, T., Saadallah, S., David, P., and Vix-Guterl, C. (2004). Hydrogen storage in activated carbon materials: role of the nanoporous texture, *Carbon*, **42**, pp. 2744–2747.

43. Gadiou, R., Saadallah, S.E., Piquero, T., David, P., Parmentier, J., and Vix-Guterl, C. (2005). The influence of textural properties on the adsorption of hydrogen on ordered nanostructured carbons, *Microporous Mesoporous Mater.*, **79**, pp. 121–128.

44. Jordá-Beneyto, M., Lozano-Castelló, D., Suárez-García, F., Cazorla-Amorós, D., and Linares-Solano, A. (2008). Advanced activated carbon monoliths and activated carbons for hydrogen storage, *Microporous Mesoporous Mater.*, **112**, pp. 235–242.

45. Zhou, L., Zhou, Y.P., and Sun, Y. (2004). Enhanced storage of hydrogen at the temperature of liquid nitrogen, *Intl. J. Hydrogen Energy*, **29**, pp. 319–322.

Chapter 4

Biomass-Derived Carbon Electrodes for Electrochemical Double-Layer Capacitors

Thomas E. Rufford,[a] Erika Fiset,[b] Denisa Hulicova-Jurcakova,[b] and Zhonghua Zhu[b]

[a]*School of Mechanical and Chemical Engineering, The University of Western Australia, M050 35 Stirling Highway, Crawley, WA 6009, Australia*
[b]*Australian Institute of Bioengineering and Nanotechnology, School of Chemical Engineering, The University of Queensland, St. Lucia, Qld 4067, Australia*

Thomas.rufford@uwa.edu.au

Waste biomass from agricultural and food production processes can be used to prepare activated carbons with high surface area for energy storage in electrochemical double-layer capacitors. The activation process and biomass carbon source can be selected to influence the pore structure and heteroatom content (for example, nitrogen and oxygen) of the activated carbon product. In this chapter, we review the effect of carbon pore structure and the effects of heteroatoms on the capacitance of carbon electrodes and compare the electrochemical performance of activated carbons prepared from three biomass wastes: coffee waste, sawdust and sugarcane bagasse, popcorn and sucrose (as a refined carbon source).

Green Carbon Materials: Advances and Applications
Edited by Thomas E. Rufford, Denisa Hulicova-Jurcakova, John Zhu

ISBN 978-981-4411-13-4 (Hardcover), 978-981-4411-14-1 (eBook)
www.panstanford.com

4.1 Carbons for Electrochemical Double-Layer Capacitors

Electrochemical double-layer capacitors (EDLCs), also known as supercapacitors, are a promising energy storage technology for applications where high power density and long-cycle life are required. These applications include auxiliary power supplies in fuel cell vehicles, systems to recover energy dissipated during a train braking, consumer electronics and memory back-up systems, and energy storage systems for capturing renewable energy sources. No charge transfer reactions are involved in charge–discharge cycles of EDLCs, unlike in the operation of conventional batteries, so EDLCs can achieve stable and reversible charge cycling at fast charge rates. However, the energy density of an EDLC is a fraction of a conventional lead acid battery. To realize the benefits of EDLCs in high-power-demand applications, such as electric fuel cell vehicles, electrode materials with high energy density, approaching that of conventional batteries, are required.

Energy density of an EDLC is proportional to the cell's specific capacitance, and capacitance depends on the electrode-electrolyte interfacial area. Therefore, porous activated carbons with high surface areas are a popular electrode material for EDLCs. In addition to high surface areas, porous carbons can provide high conductivity, good corrosion resistance, high thermal stability and controllable pore structures [1–4]. Specific capacitance correlates with micropore surface area, which is the surface area of the pores up to 2 nm wide [5, 6]. However, not all the micropores in a high-surface-area activated carbon may be effective for double-layer capacitance. In H_2SO_4, carbon pores 0.7–1 nm wide produce optimum specific capacitance [5, 7, 8], but pores less than 0.5 nm wide are generally considered too narrow for forming a double layer [6]. Although mesopores (>2 nm) have low double-layer density, several studies report [7, 11, 12] that mesopores can improve EDLC performance at fast charge rates by facilitating electrolyte diffusion through the carbon network to micropore adsorption sites. Introducing heteroatoms such as oxygen and nitrogen into the activated carbon structure can enhance specific capacitance [14]. Heteroatoms can improve carbon wettability to the electrolyte, speed charge transfer through the electrode, and, in certain functional groups, provide pseudo-Faradaic capacitance.

Although activated carbons are available at relatively low cost, the high-performance carbons currently used in commercial devices remain one of the major cost contributors in the cells [16]. Waste biomass from agricultural and food production processes has the potential to be used as a low-cost, renewable source of carbon for production of high-performance electrodes for supercapacitor devices. Furthermore, the nitrogen and oxygen content of certain biomass materials also offers the possibility to produce carbon electrodes with high concentrations of electrochemically active surface functional groups. The results in Table 4.1 show the potential of plant wastes as raw materials for activated carbons with good electrochemical properties. Supercapacitor electrodes have been prepared from biomass wastes, including seaweed biopolymers [9], corn [10], banana fibres [1] and coffee grounds [13]. The seaweed biopolymer carbon was prepared at a low temperature (873 K) under argon without any chemical activation agent; it achieves a good specific capacitance due to the pseudocapacitance of oxygen surface functionalities [9]. The corn grain was reported to contain significant concentrations of electrochemically active oxygen surface groups (quinone) [10].

Table 4.1 Survey of activated carbons from biomass precursors for electrode materials in electrochemical double layer capacitors

Carbon source	Activation method	S_{BET} (m^2/g)	Capacitance (F/g)	Electrolyte
Banana fibres [1]	$ZnCl_2$	1097	296	1 M Na_2SO_4
Banana fibres [1]	KOH	686	264	1 M Na_2SO_4
Fir wood [2]	KOH	1064	180	0.5 M H_2SO_4
Fir wood [2]	Steam	1016	110	0.5 M H_2SO_4
Seaweed biopolymer [6, 9]	Thermal	273	198	1 M H_2SO_4
Corn grain [7, 8, 10]	KOH	3199	257	6 M KOH
Coffee grounds [13]	$ZnCl_2$	1019	368	1 M H_2SO_4
Rubber wood saw dust [15]	CO_2	912	138	1 M H_2SO_4

Note: Capacitance reported here is the three-electrode equivalent for a single electrode.

The choice of carbon activation process has a strong influence on both the carbon properties and the cost of production. Clearly, the choice of carbon precursor will have a direct impact on the cost to produce the activated carbon; the precursor also can influence the carbon properties which affect EDLC performance. For example, different plant materials can supply oxygen and nitrogen atoms for pseudocapacitance, and the cellular structure of the plant wastes can influence the macroporous structure of the activated carbon.

4.2 Effects of Carbon Properties on Capacitance

4.2.1 Effect of Pore Structure

The specific capacitance of a porous electrode material depends on the electrolyte/electrode interfacial area available for the double-layer formation. Published results confirm that in general, specific capacitance increases with specific surface area. For example, Lota and co-authors [17] studied the influence of specific surface area on the specific capacitance of KOH treated activated carbons and reported that although specific capacitance in general increases the capacitance values were found to deviate from the linear best fit, due to pore size effects and the pseudocapacitance of surface functional groups, a process which is described in Section 4.2.2.

For aqueous electrolytes, it is generally considered that pores greater than 0.5 nm wide are available for double-layer formation [4]. Lin et al. reported [6] that the optimum carbon pore size for EDLC in a H_2SO_4 electrolyte is 0.8 to 2.0 nm. Raymundo-Piñero et al. report [8] maximum volumetric capacitance at a slightly narrower pore size of 0.7 nm for acidic electrolyte. Vix-Guterl [7] reported maximum volumetric specific capacitance for pores 0.7 nm wide. The solvated sulphate ion ($SO_4^{2-} \cdot 12H_2O$) has a diameter of 0.53 nm [18], and Lota et al. reported that pores as narrow as 0.6 nm do not hinder the mobility of the solvated sulphate ion [17]. Furthermore, Chmiola et al. [5, 19] proposed that desolvation of ions can give high-density double-layer formation in narrow micropores, which allows double-layer formation in pores less than 1 nm wide by large ions of organic electrolytes (such as $TEA^+ \cdot 7$(acetonitrile) 1.30 nm and $BF_4^- \cdot 9$(acetonitrile) 1.16 nm [20]).

At high charge rates, the formation of the electrochemical double layer will be restricted by the rate at which electrolyte ions can reach the electrode/electrolyte interface, and in such cases, the electrolyte ions may be restricted in entering the small pores. Mesopores are reported to facilitate electrolyte transport at fast charging rates [7, 11, 12]. The contribution of mesopore surface areas to capacitance at high current loads, up to 10 A/g, has been investigated and correlated by Rufford et al. [21].

Most industrial activated carbon processes, such as physical activation with steam or CO_2, produce microporous carbons with broad pore size distributions. Chemical activation methods can provide routes to alternative carbon pore structures than the physical activation methods. For example, alkali activation with NaOH or KOH produces microporous carbons with high specific surface areas (>2000 m^2 g^{-1}) [9, 10], however, highly microporous carbons often show poor EDLC performance at fast charge-discharge rates due to restricted ion transport within micropore networks. Activated carbon with mesoporous and macroporous structure can be obtained from carbon precursors which have an existing pore structure, such as corn cobs [14]. Mesopores can also be developed using non-alkali chemical activation. $ZnCl_2$ is widely studied for preparing activated carbons with high mesopore content from biomass materials, including macadamia shells [15] and corn cobs [16]. Other chemical activating agents studied to prepare carbon electrode materials include $FeCl_3$ [22] and $MgCl_2$ [23, 24].

4.2.2 Effect of Heteroatoms and Surface Functional Groups

Atoms of other elements are usually found on porous carbon materials, either incorporated in the graphene network or located at the carbon network edge sites. The edge sites, often featuring unpaired electrons, are more reactive, and so it is at these edge sites that most heteroatoms will be found in surface functional groups. Oxygen is the most prevalent heteroatom on carbon surfaces, but nitrogen, sulphur, phosphorous and other halogens can be found on carbon materials. The most common nitrogen and oxygen functional groups found on carbon surfaces are depicted in Fig. 4.1.

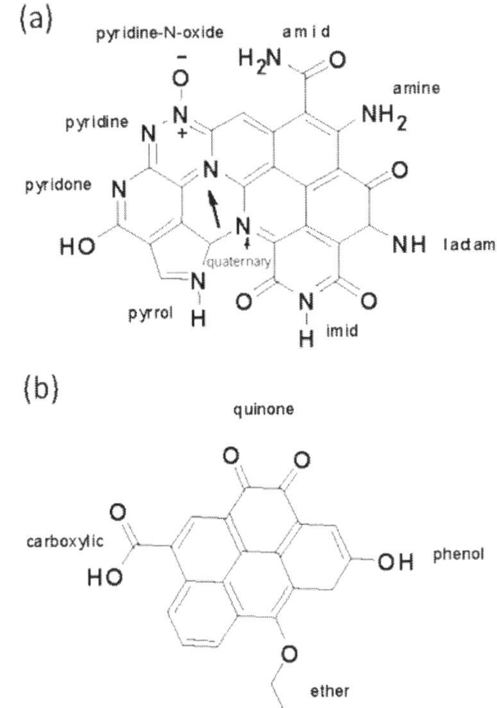

Figure 4.1 Schematic model of (a) nitrogen functionalities and (b) oxygen functionalities at the carbon surface.

Heteroatoms and surface groups can affect both the transport of the electrolyte through the carbon pores, by improving the wettability of the surface [25], the behaviour of the carbon graphene layer as an electron donor or acceptor. Thus, heteroatoms can influence the electrode performance by changing the electrical contact resistance, point of zero charge, charge adsorption, and the self-discharge characteristics [4]. Only a few surface functional groups have been found to be electrochemically active, predominantly the oxygen quinone group [26] and some nitrogen [27–30] groups. These electrochemically active groups can enhance charge storage of the electrode through pseudocapacitance.

Pseudocapacitance is an electrochemical process that shares some common characteristics with both conventional battery and EDLC processes. Pseudocapacitance can involve Faradaic charge transfer of the electroactive material as in a battery, but an electrode with pseudocapacitive properties can also exhibit linear voltage changes with current and high degrees of reversibility like an EDLC [31]. Conway et al. [31] describe some conditions that may lead to pseudocapacitance, including (1) the presence of adatom arrays on electrode surfaces (H or Cu on Pt electrodes), (2) redox processes in a liquid or solid solution where the electrode potential is a function of the log of the ratio of the reductant converted, or (3) voltage-dependent chemisorption of anions at the electrode interfaces via partial Faradaic charge transfer.

The most widely studied pseudocapacitance example is the redox reaction of the hydroquinone/quinone group [26], as shown by the reaction scheme in Fig. 4.2. Other oxygen functional groups (including hydroxyl, carbonyl, carboxyl, lactone, and pyrone groups) are present on typical activated carbons; however, pseudocapacitance behaviour has not been observed for these groups. Redox reactions such as the hydroquinone/quinone reaction can enhance specific capacitance as one electron is stored per quinone oxygen atom, compared to only 0.18 electrons per carbon atom in pure double-layer capacitance [26]. The contribution of functional groups depends on the nature of the electrolyte; for instance, Andreas and Conway reported that the capacitance of the quinone group reduced by 30% when a H_2SO_4 electrolyte was titrated with NaOH from pH 0 to 14 [26]. The quinone group does not contribute to pseudocapacitance in organic electrolytes.

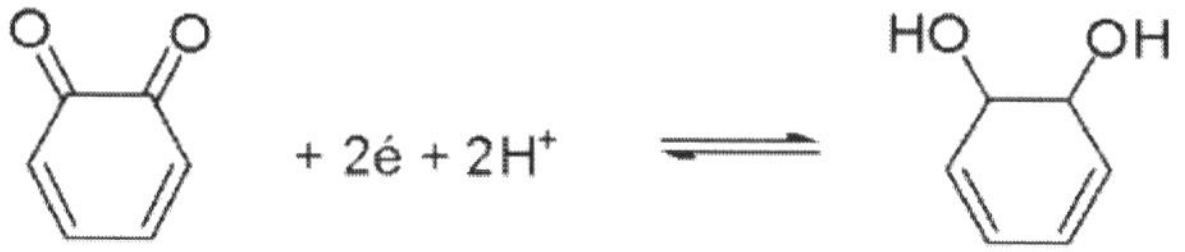

Figure 4.2 Schematic of the hydroquinone/quinone redox process.

The nitrogen functionalities (N) present on a carbon material are influenced by the chemistry of the carbon precursor and the preparation conditions. Typically carbons prepared at mild treatment temperatures (673–973 K) have N groups external to the graphene layer, such as amides, aromatic amines, and protonated amides [32]. Carbons prepared at temperatures above 973 K have fewer external functional groups, and for these high temperature carbons most N groups will be found in the aromatic rings of the graphene layer [32–34], as illustrated in Fig. 4.1. Electrochemical studies of N-enriched carbons suggest that the functional groups located at carbon edge sites (pyrollic, pyridone, and pyridinic groups) that provide pseudocapacitance [28–30, 35, 36]. The quaternary nitrogen and protonated pyridine groups may also have a positive effect on capacitance as these groups can provide delocalized charges believed to participate in Faradaic reactions [37–39].

There are three main routes to introduce nitrogen and oxygen functional groups and other heteroatoms in a carbon network [40]:

(1) Oxidative treatments to enhance surface functionalities, for example, oxidation in a mixture of air and ammonia [27].
(2) Insertion of metal oxide particles in the carbon electrode, for example, RuO_2 [41, 42] and TiO_2 [43].
(3) Carbonization and/or activation of precursor materials rich in oxygen or nitrogen. For example, seaweed biopolymer has a high oxygen content and nitrogen-rich precursors include melamine [28, 29], polyvinylpyridine and coffee beans [44]. The synthesis of activated carbon electrode materials from coffee beans, along with two other types of waste biomass, is discussed in this chapter.

4.3 Electrodes from Different Waste Biomass Sources

The influence of structure in carbon sources was investigated by $ZnCl_2$ activation of four carbon sources: waste coffee grounds, sugarcane bagasse, sawdust, and sucrose. Each of the carbon precursors investigated has a different macrostructure and chemical composition. Sucrose, unlike the biomass wastes, has no inherent cellular macrostructure as it dissolves readily in water. The sawdust (from a soft pinewood) has an open macroporous structure and cellulosic composition.

Sugarcane bagasse also has a fibrous macroporous structure and high cellulose content. Sugarcane bagasse is a by-product from the milling of sugarcane and large quantities of this waste material are produced each milling season in Australia and other sugarcane-producing countries. Activated carbons prepared from sugarcane bagasse have been studied for applications such as waste water treatment [10, 11] and gold extraction [12]. Sugarcane bagasse has an ash content of several percent weight (2.8 wt% reported by Castro et al. [45]), with a high proportion of the ash found as silica [46].

Roasted coffee beans feature open macroporous pits (20–50 nm in diameter) developed during the roasting process [47], and the beans contain a mixture of lignin-cellulosic material and starch. In 2008, the global coffee production exceeded 8 million tonnes of green coffee beans [48] (coffee is the second most traded commodity after petroleum oil) and the waste biomass produced each year in manufacturing coffee could be used to produce activated carbons for energy storage. Most of the coffee solid waste generated in instant (soluble) and espresso coffee production is disposed to land fill, in addition to millions of tonnes of husk and pulp from the primary coffee berry processing operations [49]. As spent coffee grounds have already been milled, coffee grounds may offer some carbon-manufacturing advantages over wastes such as coconut husks or tyres, which need to be shredded prior to activation. On this basis, waste coffee grounds may be a cost competitive, and renewable, carbon source for high-performance niche applications such as supercapacitor electrodes.

Green coffee beans are roasted at about 473 K to develop flavours and colour. During roasting, pyrolysis reactions create pores in the coffee beans. The pores in roasted coffee beans have typically 20–50 nm diameter, and the interconnectedness of pores increases with roasting severity (temperature) [47]. The coffee grounds reported in this chapter were collected from a domestic espresso machine on a single day. This collection procedure ensured a single blend (that is a single brand and roast of coffee) and constant particle size (fineness of grind) for all experiments reported in this thesis. The BET specific surface area of these waste coffee grounds measured by N_2 sorption analyses at 77 K was 17 m^2/g.

4.3.1 Materials and Methods

4.3.1.1 Preparation and characterization of activated carbons

The sawdust was gathered by hand sawing a piece of soft pinewood. The sugarcane bagasse was milled in a coffee grinder, roughly sieved, then the smaller size fraction was demineralized in hot water at 373 K for 8 h, filtered and dried at 373 K. Reagent-grade sucrose was used. The waste coffee grounds (*coffea arabica*) were collected from the tea-room espresso machine and dried at 373 K for 24 h.

In a typical activation procedure, 2 g of carbon precursor and 2 g of $ZnCl_2$ in 20 mL water were stirred at room temperature for

4 h. Then the mixture was dried in an oven at 373 K overnight. Carbonization was performed under N_2 gas flow in a tube furnace with a heating rate of 5 K/min to 1173 K. The maximum temperature was held for 1 h. The carbonized samples were washed in 0.2 M HCl, rinsed, filtered and then dried to obtain the activated coffee grounds carbon (CGC), pinewood sawdust carbon (PWC), sugarcane bagasse carbon (BaC), and sucrose carbon (SuC).

The surface textures and porosity of the activated carbons were examined by N_2 sorption at 77 K (Quantachrome Quadrasorb SI) and CO_2 sorption at 273 K (Quantachrome Autosorb 1-C). Specific surface areas were calculated from the N_2 adsorption isotherms using the BET equation. Pore size distributions were calculated using non-local density functional theory algorithms for carbon slit pores (Quadrawin 2.0 software package).

4.3.1.2 Electrochemical measurements

The capacitance of the activated carbons was evaluated in two-electrode sandwich type cells filled with 1 M H_2SO_4. Electrodes were prepared by coating areas of 1 cm^2 on titanium strips with a film of carbon and binder. The film was applied from a slurry of 90 wt% active carbon material (for example CGC), 5 wt% carbon black (Mitsubishi Black #32), and 5 wt% polyvinylidene-fluoride (in *N*-methyl pyrrolidone. The mass of active material on each electrode was approximately 2 mg.

Cyclic voltammetry (CV) and galvanostatic charge–discharge (GC) cycles were measured in the voltage range 0–1 V (Solartron 1480 Multistat). Specific gravimetric capacitances (C) were calculated per single electrodes as the three-electrode cell equivalent [25]. However, as per the standard convention, in the construction of the Ragone plots, the two-electrode cell capacitance (C_{2E}) values were used to calculate specific energy ($E = 0.5C_{2E}U^2$, where U is the cell voltage omitting any iR drop) and specific power ($P = 0.5IU/m$, where I is the current and m the total mass of active material in two electrodes).

4.3.2 Influence of Carbon Precursor on Activated Carbon Properties

Table 4.2 summarizes the surface textural properties of carbons from $ZnCl_2$-activated pinewood (PWC), coffee grounds (CGC), sucrose

(SuC), and bagasse (BaC). At these activation conditions (1173 K and $ZnCl_2$ ratio equal to 1), the sucrose produces carbon with the highest surface area, whilst pinewood produces the carbon with largest pore volume.

Table 4.2 Textural properties of activated carbons from coffee grounds, sawdust, bagasse, popcorn and sucrose precursors activated at 1173 K using a $ZnCl_2$ to precursors mass ratio of 1

Activated carbon	Precursor	S_{BET} (m^2/g)	V_{total} (cm^3/g)	V_{micro}[a] (cm^3/g)
CGC	Coffee grounds	977	0.48	0.39
PWC	Pinewood	1262	0.72	0.60
BaC	Sugarcane bagasse	1155	0.64	0.38
SuC	Sucrose	1434	0.66	0.59

[a]Cumulative volumes of pores narrower than 2 nm from the N_2 sorption DFT analyses.

Figure 4.4 shows the N_2 isotherm for PWC that slopes continuously with increasing relative pressure, and this isotherm shape is typical of a wide mesopore and macropore distribution. The open cellular structures of pinewood and bagasse produce an activated carbon skeleton with open mesopores and macropores, as seen the in the pore size distributions of Fig. 4.3b. In contrast, the

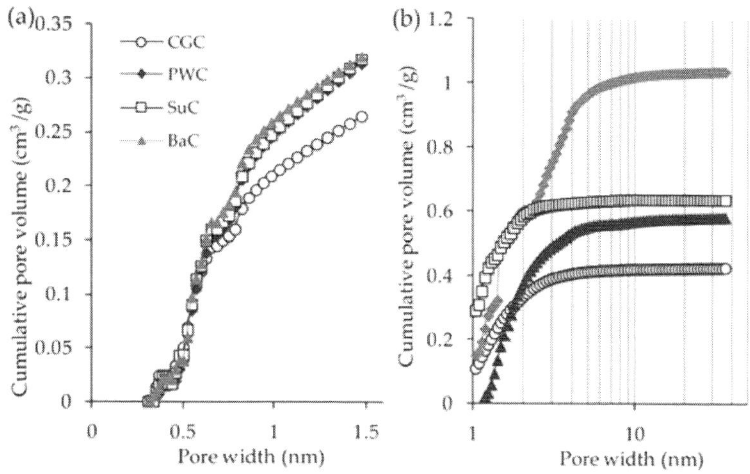

Figure 4.3 Cumulative pore size distribution (DFT) in activated carbons from coffee grounds (CGC), sawdust (PWC), sucrose (SuC) and sugarcane bagasse (BaC) calculated from (a) CO_2 at 273 K and (b) N_2 at 77 K sorption isotherms.

N_2 isotherm of SuC is clearly Type I [50] and has the narrowest pore size distribution for pores 1–3 nm. The dissolved sucrose presents no inherent macrostructure and when activated with $ZnCl_2$ produced a carbon with a high degree of microporosity only. The homogeneous distribution of the $ZnCl_2$ through the sucrose allows a more uniform pore size distribution, as seen in Fig. 4.3b, and higher surface area.

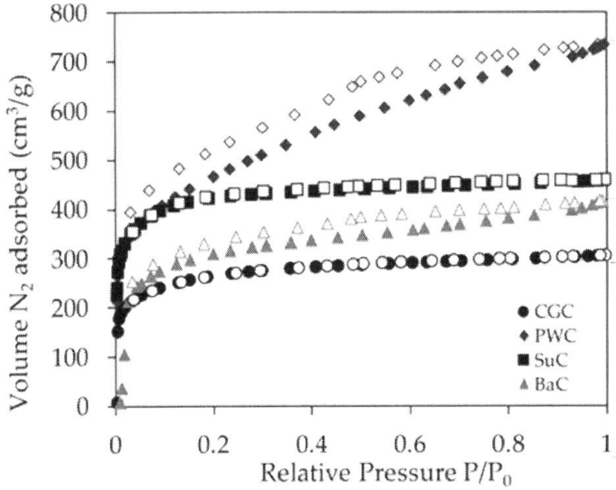

Figure 4.4 N_2 adsorption/desorption isotherms (closed/open symbols) at 77 K on $ZnCl_2$ activated carbons from coffee grounds, pinewood, sucrose and sugarcane bagasse.

4.3.3 Electrochemical Performance

The amplitude of the CV curves in Fig. 4.5 show that each of the activated carbons exhibits good EDLC performance in H_2SO_4. The rectangular CV shapes are typical of carbons with good electrical conductivity. At 100 mV/s the PWC and SuC retain the squarest CV curves. The GC profiles (not shown) for the four $ZnCl_2$-activated carbons, were symmetrical at charge–discharge current loads from 0.05 to 5 A/g. Symmetrical GC profiles are typical of reversible EDLC behaviour. At high current loads (5 A/g) small voltage drops (*iR* drops), due to internal resistances of cell, are seen in the GC profiles. The *iR* drops of CGC, PWC, SuC, and BaC at 5 A/g are only 0.03 V, which provides further evidence of good EDLC behaviour at fast charge rates.

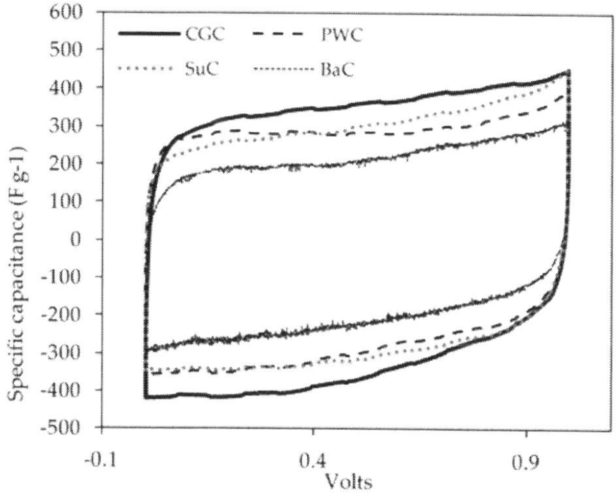

Figure 4.5 Cyclic voltammetry profiles of $ZnCl_2$ activated carbons prepared from coffee grounds (CGC), pinewood sawdust (PWC), sucrose (SuC) and sugarcane bagasse (BaC), measured at scan rates of 5 mV/s. Specific capacitance here is calculated for a single electrode, using the equation $C = 4i\Delta U/\Delta t$, from CV in a two-electrode cell.

Specific capacitance (C) of the carbon electrodes was calculated from the discharge side of GC profiles. At low current loads, CGC has the highest specific capacitance, although all four carbon samples achieve specific capacitances greater than 200 F/g at current loads less than 0.1 A/g. A specific capacitance above 200 F/g is excellent for activated carbon in an aqueous electrolyte [4, 14]. Figure 4.6 shows specific capacitance drops with increasing current load, which is a trend typical for all porous electrodes, and at current loads approaching 10 A/g the PWC and SuC-derived carbons show the best capacitance.

From the electrochemical results presented in Figs. 4.5 and 4.6, two key differences between the samples can be observed: (1) The bagasse carbon has a similar specific surface area to the pinewood carbon and a similar total pore volume to the sucrose-derived carbon; yet the capacitance of the BaC electrodes is significantly lower than that of the other carbons, and (2) the coffee ground carbon achieves a better capacitance per unit of surface area (or pore volume) than the other carbons. We have discussed the performance of the CGC in detail elsewhere [13, 44]. In brief, we attribute the excellent electrochemical performance of the CGC to a well developed pore structure with a large percentage of optimum sized pores for double-layer formation in H_2SO_4, complemented by pseudo-Faradaic reactions involving oxygen and nitrogen functional groups. The electrochemical active functional groups provide only a minor contribution, estimated at 8–10% of the total capacitance.

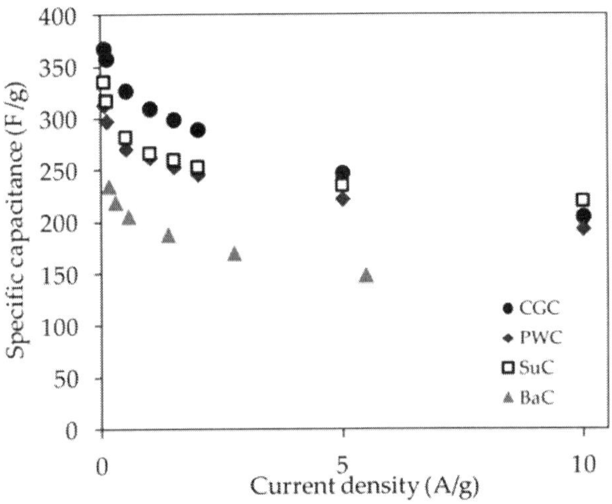

Figure 4.6 The effect of current density on specific capacitance in 1 M H_2SO_4 of electrodes prepared from $ZnCl_2$ activated carbons. Specific capacitance in this figure is the three-electrode equivalent capacitance of a single electrode calculated from the discharge side of GC measurements in a two-electrode cell.

The lower capacitance of the BaC, compared with the other carbons studied here, is attributed to a lower micropore volume than the SuC and PWC and the presence of impurities in BaC. The optimum pore size for double-layer capacitance in H_2SO_4 is reported to be in the range 0.5–2 nm diameter [5, 6, 51] and, as shown in Table 4.2, the BaC has a lower micropore volume than the SuC and the PWC.

Sugarcane bagasse has a high ash content, including high silica content, [45, 46] which can result in high levels of ash in the carbon product. The estimates of ash analysis results in Table 4.3 confirm that the bagasse used in this study has high ash content.

Given the ash content of the bagasse, the 3.9 wt% ash in BaC is, however, unexpectedly low. The method to estimate ash content was combustion in air at 1023 K using TGA, and the error here may be greater than the method describe in Australian Standard AS1038.3 (*Coal and coke—Analysis and testing, Part 3: Proximate analysis of higher rank coal*). XPS analysis shows peaks in the wide survey spectra at low binding energies (around 155.6 and 106.4 eV), which we attribute to Si and Al impurities.

Table 4.3 Estimated moisture and ash content of raw materials and activated carbons from coffee grounds and sugarcane bagasse

Precursor	Moisture content (wt%)	Ash content (wt%)
Coffee grounds	2.3	1.7
Sugarcane bagasse	5.4	10.5
Coffee ground carbon (CGC)	16.3	3.1
Bagasse carbon (BaC)	8.3	3.9

In Fig. 4.7 the negative impact of impurities on the electrochemical performance of the BaC electrodes is further highlighted by the loss of capacitance during charge-discharge cycling. The PWC, CGC and SuC show a drop in capacitance of 5–7% after 5000 cycles at a fast charge-discharge rate of 5 A/g, where the BaC capacitance drops by more than 20%. Impurities in the carbon electrodes can promote degradation of the electrolyte, and the other carbons should have lower level of impurities than the BaC. Estimated ash in waste coffee grounds was only 1.7 wt% (Table 4.3), which is much lower than the bagasse. No significant XPS peaks associated with impurities were observed for activated carbon CGC. Although we have not performed XPS or TGA for SuC or PWC, we expect no Si or Al to be present in the sucrose and very little of these elements in the pinewood. These results suggest that the bagasse demineralization (water at 373 K) and carbon rinsing (0.2 M HCl at room temperature) procedures do not adequately remove silica or alumina from the bagasse and BaC. Further optimization of this bagasse activation procedure is required to produce a better BaC for electrode applications.

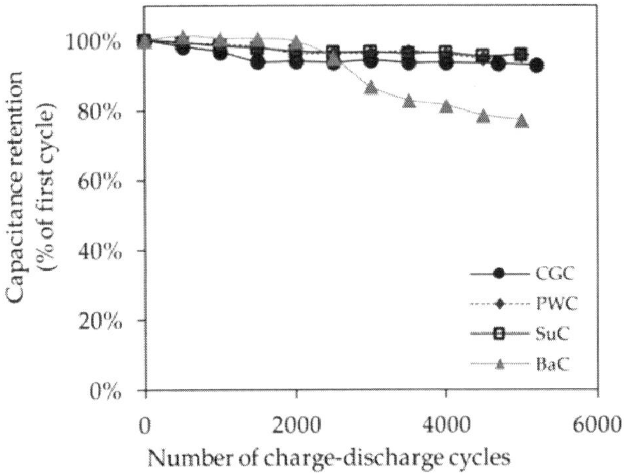

Figure 4.7 Electrochemical stability of supercapacitor electrodes made from $ZnCl_2$ activated coffee grounds, pinewood sawdust, sucrose and sugarcane bagasse in galvanostatic cycling at 5 A/g.

The excellent performance of the CGC, PWC and SuC carbons is demonstrated in the Ragone plot of Fig. 4.8. The performance of our $ZnCl_2$-activated carbons is compared to biomass-derived carbons reported in the literature, including a seaweed by-product [9] and $ZnCl_2$-activated banana fibres (BFC) [1]. Subramanian et al. [1] reported capacitance for only one current load for the BFC in Na_2SO_4 electrolyte, and the specific energy of this data point is close to the curves for SuC and SDC. As the electrolyte anion is the same as H_2SO_4, we could expect that BFCs would present similar Ragone plots to SDC and SuC over a range of current loads. However, retention of performance at high power density would be strongly influenced by diffusion of ions through mesopores. Banana fibres have a open pore structure [1] and, as the BFC was prepared with a low $ZnCl_2$ ratio of 0.1, which produces micropores, it will be the plant macrostructure that contributes mesopores (and macropores) to the carbon.

Organic electrolyte based supercapacitors can achieve higher energy densities than aqueous electrolytes, as energy density is proportional to the square of cell voltage and organic electrolytes can typically be operated at cell potentials of up to 2.7 V. Due to the larger size of organic salt ions, such as the ions of tetraethylammonium tetrafluoro borate ($TEABF_4$) the optimum pore size for double-layer capacitance in organic electrolytes is larger than for H_2SO_4. The carbon preparation procedure used in the current work can be readily tailored to produce activated carbons with larger mesopores, by control of the $ZnCl_2$ ratio. We have obtained specific energy of up to 34 W/kg (based on the mass of active electrode material) in 1 M $TEABF_4$/acetonitrile for activated carbons from waste coffee grounds [44], which provides evidence that $ZnCl_2$-activated carbons from waste biomass have potential for a wide range of energy and power demand applications.

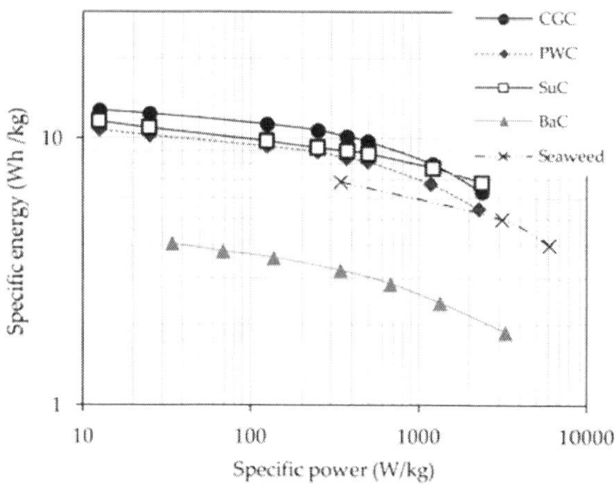

Figure 4.8 Ragone plot for supercapacitors with electrodes prepared from coffee grounds, pinewood sawdust, sucrose and sugarcane bagasse. Specific energy and specific power in this plot are based on the mass of active electrode material, and exclude the mass of the electrolyte, current collectors, and cell packaging.

4.4 Conclusions

Activated carbons prepared from pinewood sawdust, sugarcane bagasse, sucrose and waste coffee grounds can achieve high specific capacitance in 1 M H_2SO_4. Activated carbons with surface areas from 977 to 1434 m^2/g were obtained. The properties of the carbon precursors influence the mesopore and macropore structure, surface functional groups, and purity of the activated carbon products. However, the $ZnCl_2$ activation dominates the development of the pore structure and hence all precursors presented activated carbons with good electrochemical properties. High ash content in the BaC reduced the specific capacitance of this sample and led to poor charge cycling stability.

References

1. Subramanian, V., *et al.*, Supercapacitors from activated carbon derived from banana fibers. *J. Phys. Chem. C*, 2007. **111**: 7527–7531.
2. Wu, F.-C., *et al.*, Effects of pore structure and electrolyte on the capacitive characteristics of steam- and KOH-activated carbons for supercapcitors. *J. Power Sources*, 2005. **144**: 302–309.
3. Kötz, R., and Carlen, M., Principles and applications of electrochemical capacitors. *Electrochim. Acta*, 2000. **45**: 2483–2498.
4. Pandolfo, A.G., and Hollenkamp, A.F., Carbon properties and their role in supercapacitors. *J. Power Sources*, 2006. **157**: 11–27.
5. Chmiola, J., *et al.*, Effect of pore size and surface area of carbide derived carbons on specific capacitance. *J. Power Sources*, 2006. **158**: 765–772.
6. Lin, C., Ritter, J.A., and Popov, B.N., Correlation of double-layer capacitance with the pore structure of sol-gel derived carbon xerogels. *J. Electrochem. Soc.*, 1999. **146**(10): 3639–3643.
7. Vix-Guterl, C., *et al.*, Electrochemical energy storage in ordered porous carbon materials. *Carbon*, 2005. **43**(6): 1293–1302.
8. Raymundo-Piñero, E., *et al.*, Relationship between the nanoporous texture of activated carbons and their capacitance properties in different electrolytes. *Carbon*, 2006. **44**(12): 2498–2507.
9. Raymundo-Piñero, E. Leroux, F., and Béguin, F., A high-performance carbon for supercapacitors obtained by carbonization of a seaweed biopolymer. *Adv. Mater.*, 2006. **18**: 1877–1882.
10. Balathanigaimani, M.S., *et al.*, Highly porous electrodes from novel corn grains-based activated carbons for electrical double layer capacitors. *Electrochem. Commun.*, 2008. **10**: 868–871.
11. Wang, D.-W., *et al.*, 3D aperiodic hierachical porous graphitic carbon material for high-rate electrochemical capacitive energy storage. *Angew. Chem.-Int. Ed.*, 2008. **47**: 373–376.
12. Xing, W., *et al.*, Superior electric double layer capacitors using ordered mesoporous carbons. *Carbon*, 2006. **44**(2): 216–224.

13. Rufford, T.E., *et al.*, Nanoporous carbon electrode from waste coffee beans for high performance supercapacitors. *Electrochem. Commun.*, 2008. **10**: 1594–1597.

14. Frackowiak, E., Carbon materials for supercapacitor application. *Phys. Chem. Chem. Phys.*, 2007. **9**: 1774–1785.

15. Taer, E., *et al.*, Preparation of a highly porous binderless activated carbon monolith from rubber wood sawdust by a multi-step activation process for application in supercapcitors. *Intl. J. Electrochem. Sci.*, 2011. **6**(8): 3301–3315.

16. Burke, A., R&D considerations for the performance and application of electrochemical capacitors. *Electrochim. Acta*, 2007. **53**: 1083–1091.

17. Lota, G., *et al.*, Improvement of the structural and chemical properties of a commercial activated carbon for its application in electrochemical capacitors. *Electrochim. Acta*, 2008. **53**: 2210–2216.

18. Endo, M., *et al.*, Capacitance and pore-size distribution in aqueous and nonaqueous electrolytes using various activated carbons. *J. Electrochem. Soc.*, 2001. **148**(8): A910–A914.

19. Chmiola, J., *et al.*, Desolvation of ions in subnanometer pores and its effect on capacitance and double-layer theory. *Angew. Chem.-Int. Ed.*, 2007. **47**: 3392–3395.

20. Kim, Y.-J., *et al.*, PVDC-based carbon material by chemical activation and its application to nonaqueous electrolytes. *J. Electrochem. Soc.*, 2004. **151**(6): E199–E205.

21. Rufford, T.E., *et al.*, Empirical analysis of the contributions of mesopores and micropores to the double-layer capacitance of carbons. *J. Phys. Chem. C*, 2009. **113**: 19335–19343.

22. Oliveira, L.C.A., *et al.*, Preparation of activated carbons from coffee husks utilizing $FeCl_3$ and $ZnCl_2$ as activating agents. *J. Hazard. Mater.*, 2009. **165**(1–3), pp. 87–94.

23. Rufford, T.E., *et al.*, A comparative study of chemical treatment of $FeCl_3$, $MgCl_2$ and $ZnCl_2$ on microstructure, surface chemistry and double-layer capacitance of carbons from waste biomass. *J. Mater. Res.*, 2010. **8**: 1451–1459.

24. Morishita, T., *et al.*, Preparation of porous carbons from thermoplastic precursors and their performance for electric double layer capacitors. *Carbon*, 2006. **44**(12): 2360–2367.

25. Qu, D., Studies of activated carbons used in double-layer supercapacitors. *J. Power Sources*, 2002. **109**(2): 403–411.

26. Andreas, H.A., and Conway, B.E., Examination of the double-layer capacitance of an high specific-area C-cloth electrode as titrated from acidic to alkaline pHs. *Electrochim. Acta*, 2006. **51**: 6510–6520.
27. Jurewicz, K., *et al.*, Ammoxidation of brown coals for supercapacitors. *Fuel Process. Technol.*, 2002. **77–78**: 191–198.
28. Hulicova, D., Kodama, M., and Hatori, H., Electrochemical performance of nitrogen-enriched carbons in aqueous and non-aqueous supercapacitors. *Chem. Mater.*, 2006. **18**(9): 2318–2326.
29. Hulicova, D., *et al.*, Supercapacitors prepared from melamin-based carbon. *Chem. Mater.*, 2005. **17**(5): 1241–1247.
30. Frackowiak, E., *et al.*, Optimisation of supercapacitors using carbons with controlled nanotexture and nitrogen content. *Electrochim. Acta*, 2006. **51**(11): 2209–2214.
31. Conway, B.E., Birss, V., and Wojtowicz, J., The role and utilization of pseudocapacitance for energy storage of supercapacitors. *J. Power Sources*, 1997. **66**(1): 1–14.
32. Seredych, M., *et al.*, Surface functional groups of carbons and the effects of their chemical character, density and accessibility to ions on electrochemical performance. *Carbon*, 2008. **46**(11): 1475–1488.
33. Mangun, C.L., *et al.*, Surface chemistry, pore sizes and adsorption properties of activated carbon fibers and precursors treated with ammonia. *Carbon*, 2001. **39**(12): 1809–1820.
34. Pels, J.R., *et al.*, Evolution of nitrogen functionalities in carbonaceous materials during pyrolysis. *Carbon*, 1995. **33**(11): 1641–1653.
35. Lota, G., *et al.*, Effect of nitrogen in carbon electrode on the supercapacitor performance. *Chem. Phys. Lett.*, 2005. **404**(1–3): 53–58.
36. Ania, C.O., *et al.*, The large electrochemical capacitance of microporous doped carbon obtained by using a zeolite template. *Adv. Funct. Mater.*, 2006. **17**(11): 1828–1836.
37. Ingram, M.D., *et al.*, Development of electrochemical capacitors incorporating processable polymer gel electrolytes. *Electrochim. Acta*, 1998. **43**(10–11): 1601–1605.
38. Hulicova-Jurcakova, D., *et al.*, Combined effect of nitrogen- and oxygen-containing functional groups of microporous activated carbon on its electrochemical performance in supercapacitors. *Adv. Funct. Mater.*, 2009. **19**(3): 438–447.
39. Hulicova-Jurcakova, D., *et al.*, Nitrogen-enriched nonporous carbon electrodes with extraordinary supercapacitance. *Adv. Funct. Mater.*, 2009. **19**(11): 1800–1809.

40. Frackowiak, E., and Béguin, F., Carbon materials for the electrochemical storage of energy in capacitors. *Carbon*, 2001. **39**: 937–950.

41. Fang, W.-C., *et al.*, Arrayed CN_xNT-RuO_2 nanocomposites directly grown on Ti-buffered Si substrate for supercapacitor applications. *Electrochem. Commun.*, 2007. **9**(2).

42. Qin, X., Durbach, S., and Wu, G.T., Electrochemical characterization on RuO_2-H_2O/carbon nanotubes composite electrodes for high energy density supercapacitors. *Carbon*, 2004. **42**(2): 451–453.

43. Liang, H., *et al.*, Electrochemical study of activated carbon-semiconducting oxide composites as electrode materials of double-layer capacitors. *Electrochim. Acta*, 2004. **49**(21): 3463–3467.

44. Rufford, T.E., *et al.*, Double-layer capacitance of waste coffee ground activated carbons in an organic electrolyte. *Electrochem. Commun.*, 2009. **11**: 974–977.

45. Castro, J.B., *et al.*, Phosphoric acid activation of agricultural residues and bagasse from sugar cane: inlfuence of the experimental conditions on adsorption characteristics of activated carbons. *Ind. Eng. Chem. Res.*, 2000. **39**(11): 4166–4172.

46. Tsai, W.T., *et al.*, Characterization of activated carbons prepared from sugarcane bagasse by $ZnCl_2$ activation. *J. Environ. Sci. Health B*, 2001. **B36**(3): 365–378.

47. Schenker, S., *et al.*, Pore structure of coffee beans affected by roasting conditions. *J. Food Sci.*, 2000. **65**(3): 452–457.

48. Nations, F.a.A.O.o.t.U. *FAOSTAT*. 2010 [cited 2010 09-Feb-2010]; Available from: http://faostat.org.

49. Oliveira, L.C.A., *et al.*, Preparation of activated carbons from coffee husks utilizing $FeCl_3$ and $ZnCl_2$ as activating agents. *J. Hazard. Mater.*, 2008. **165**(1–3): 87–94.

50. Rouquerol, J., *et al.*, IUPAC recommendations for the characterization of porous solids. *Pure Appl. Chem.*, 1994. **66**(8): 1739–1758.

51. Lozano-Castello, D., *et al.*, Influence of pore structure and surface chemistry on electric double layer capacitance in non-aqueous electrolyte. *Carbon*, 2003. **41**(9): 1765–1775.

Chapter 5

Cellulose-Based Nanostructured Carbons for Energy Conversion and Storage Devices

Joanna Rooke,[a] Romain Sescousse,[b] Tatiana Budtova,[b] Sandrine Berthon-Fabry,[c] Bernard Simon,[d] and Marian Chatenet[a]

[a]*Laboratoire d'Electrochimie et de Physicochimie des Matériaux et des Interfaces, LEPMI, UMR 5279, CNRS/Grenoble-INP/UdS/UJF, 1130 rue de la piscine, BP75, 38402 Saint Martin d'Hères Cedex, France*

[b]*Mines ParisTech, Centre de Mise en Forme des Matériaux, BP 207, 06904, Sophia-Antipolis Cedex, France*

[c]*Mines ParisTech, Centre Énergétique et Procédés, BP 207, 06904, Sophia-Antipolis Cedex, France*

[d]*Service Recherche, Saft, 111–113, boulevard Alfred Daney, 33074, Bordeaux Cedex, France*

Marian.Chatenet@phelma.grenoble-inp.fr

New nanostructured carbons based on cellulose-derived aerogel (carbon aero cellulose [CAC]) have been developed as a "greener", biomass-based replacement for carbon black (CB) in materials used for energy storage and conversion. The correlation between the formulation, synthesis conditions, material morphology and electrochemical properties is presented and discussed. Two examples in which CAC can be used—proton exchange membrane fuel cell (PEMFC) and primary battery ($Li/SOCl_2$)—are given. We

Green Carbon Materials: Advances and Applications
Edited by Thomas E. Rufford, Denisa Hulicova-Jurcakova, John Zhu

ISBN 978-981-4411-13-4 (Hardcover), 978-981-4411-14-1 (eBook)
www.panstanford.com

show that by tailoring the texture of these new "green" carbons, we can increase the electrochemical performance in comparison with carbon black (CB).

5.1 Introduction

For economic and environmental reasons, there is a growing demand for innovative materials, devices and systems that are efficient at energy saving, storage and conversion. Recent political achievements (Kyoto Protocol, Reach Directive, etc.) have raised public awareness and driven the market need for systems that use energy efficiently yet have a low environmental impact. Obviously an electrochemical generator, which intrinsically displays a higher fuel efficiency than an internal combustion engine (whilst not emitting pollutants such as CO_x, NO_x, SO_x) or a high charge/discharge efficiency are particularly attractive. Nevertheless, although the market offers numerous possibilities in terms of electrochemical generators, further innovations with regard to their constituent parts, in particular their electrodes, is mandatory to increase their performance and operation efficiency and most of all to limit their environmental impact.

Carbon materials are widely used as electrode materials in electrochemical generators and storage devices owing to their easily tuned physicochemical properties [1–6]. Two allotropes of carbon with well-defined properties are diamond (sp^3) and graphite (sp^2); however, numerous hybrid forms exist, depending on the manufacturing process employed. They are grouped in different families such as carbon blacks (CB) [7], activated carbons (AC) [8], carbon nanotubes (CNT) [9], fullerenes (F) [10], mesoporous carbons (MC) [11], carbon aerogels (CA) [12] and carbon xerogels (CX) [13]. When a carbon is used in electrochemical energy storage and conversion devices, it must possess certain characteristics suited to such applications. Composition (surface and bulk) and structure (textural properties such as specific surface area, porous volumes and pore size distribution) often need to be adjusted on multiple length scales, from the macro down to the nanoscale. Such carbons should possess high electrical conductivity (e.g. graphite pellets are used in the positive electrode of Li-ion batteries for electron percolation), be chemical inert (they should be resistant to corrosion, which can be induced by the electrode potential or the presence of corrosive media), lightweight and mechanically resistant with minimal impurities that could inhibit the electrode activity [14]. They should also have a tailored porous network to enable mass transport throughout a bulk electrode (e.g. for electrolyte intrusion in batteries and electrolyte/reactant (gases or liquid) intrusion/mass transport in PEMFC electrodes) [15].

The efficiency and performance of the resultant electrodes, where the electrochemical reaction occurs, strongly depend not only on the characteristics of the carbon support but also on the other electrode constituent binder (e.g. in battery), electrical conductivity agent (e.g. Li-ion cathode), (electro)catalyst (e.g. in PEMFC or certain types of supercapacitor), or proton-conductive polymer (in PEMFC)—and the way they are implemented [16–17].

In addition, electrochemical devices such as PEMFCs and batteries often rely on the use of metals such as platinum and lithium, respectively, which are expensive and scarce [18]. In order to minimise the overall electrode cost, different strategies are followed according to the electrochemical device. In PEMFCs, the (usually noble metal based) electrocatalysts are not used in bulk form, as this proves ineffective in terms of cost-to-performance ratio. Instead, they are dispersed as very small particles (nanoscale) over the well-tailored carbon support (texture and chemistry), which not only reduces the loading of the metal required but also improves its utilisation factor (u_{Pt}) and effectiveness factor (ε) [19]. While u_{Pt} requires only maximal metal dispersion and both electronic percolation to the current collector and ionic percolation to the ion-exchange membrane, ε can only be optimised provided the porous network of the electrode favours a high rate of mass transport of the reactants/products; in other words, the electrocatalyst can only operate efficiently provided the interplay between the three different phases (carbon support, ionomer and porous network) is optimised. This emphasises the importance of the carbon support in terms of physical properties (electron conductivity, surface area, porous texture, degree of organisation of the carbon framework). In primary batteries, although the system is intrinsically simpler because only two networks are interpenetrated (the electron-conductive electrode and the ion-conductive electrolyte), the porosity of the electrode must be adapted to the size of the LiCl deposit that forms on the carbon surface during discharge, to avoid prematurely blocking the pores, which would ultimately limit the lithium utilisation rate.

Traditionally carbons such as CB are used to support electrocatalysts in PEMFCs or as an electrically conductive additive in batteries, owing to their high surface area to volume ratios. Depending on the fabrication method, the grain size of CBs can vary from small particles of a few nanometres with high BET specific areas to particles of around 50 nm (e.g. 250 m^2 g^{-1} for Vulcan XC72, one of the most widely used supports for PEMFC electrocatalysts). In such cases, the textural properties of the electrode, pore size and specific surface area, cannot be tailored independently [14, 20–22]. Their natural aggregation *via* van der Waals forces is also a problem, as this favours a facile rearrangement of the porous structure of the electrode during operation [17]. Furthermore, CBs originate from the incomplete combustion of heavy petroleum feedstocks and can therefore not be considered as green materials in contrast to biomass-derived carbons.

Unlike CBs, CAs or aerogel-like materials possess a 3-D random close packed structure [23, 24] comprising covalently bonded carbon spheres (3–30 nm in diameter), which overlap slightly creating a monolithic porous structure with micro-, meso- and macroporosities, depending on the synthesis parameters. CAs are unique materials, owing to their high electrical conductivities, [25, 26] tailored porosities [27, 28] and high surface areas. Obtained from the pyrolysis of organic aerogels, CA were first synthesised in 1989 by Pekala *via* the polycondensation of resorcinol or melamine with formaldehyde [13, 25, 29]. These materials have subsequently been used with some success as tailored support materials for nano-electrocatalyts in PEMFC electrodes [30–39]. However, these successes were based on the use of CA from the resorcinol-formaldehyde (RF) family, which like CBs cannot be considered as green materials.

Considering that there is an immediate need for nanostructured carbon materials with a tailored porous texture that have minimal impact on the environment (i.e. so-called green materials), researchers are actively developing novel green CAs or aerogel-like materials. More environmental-friendly nanostructured carbons can be prepared from a variety of biomass-derived gels using chemical (i.e. sol-gel) or physical methods. However, the work is still in its infancy with only a few carbons having been produced to date. *Via* sol-gel technology, examples include cellulose-acetate [40–43], hemi-cellulose citrate-chitosan [44] and tannin-formaldehyde [45], whereas *via* physical methods, carbons have been derived from cellulose [46–48], lignin [49], xylane [50, 51], alginate [52]), starch [53] or chitin [54]. Fewer still have been both pyrolysed and evaluated in electrochemical devices [42, 48, 55–58].

Herein we present the fabrication of CAs derived from pure cellulose and the evaluation of these nanostructured carbons as electrode materials in PEMFCs and lithium/$SOCl_2$ primary batteries.

5.2 Carbon Aerocellulose: Preparation and Morphology

CAs derived from cellulose can be fabricated through the pyrolysis of aerocellulose, an aerogel-like material based on pure cellulose. The details of the preparation and morphological properties of this material can be found in the literature [47, 59–62]. The preparation of aerocellulose [47] is inspired by the fabrication of aerogels (Fig. 5.1). First, cellulose is dissolved until a homogeneous solution is obtained. The cellulose solution may form a gel, depending on the solvent used; however, both gel and solution can be employed. When using carbon aerocellulose (CAC) in primary batteries, where the shape of the carbon is important, a gel is required. A gel can be prepared by dissolving cellulose in 8 wt% aqueous solution of NaOH. At this concentration the mixture gels at room temperature in a few hours [63]. The shape of the mould in which the sample is gelled is maintained throughout the process and at the end a pyrolysed aerocellulose monolith is obtained. After dissolution, the cellulose undergoes coagulation (this process is often called "regeneration"). The sample, in this case the gel that is formed from cellulose dissolved in 8 wt% NaOH(aq), is placed in a non-solvent bath, often water or ethanol or a mixture of both. The solvent in which the cellulose was dissolved diffuses from the gel into the bath and is gradually replaced by the non-solvent. As a result a block of coagulated cellulose with non-solvent in the pores is formed. To maintain the porosity of the regenerated cellulose in the dry aerocellulose state (particularly the small mesopores and micropores), drying must be done under super-critical conditions using CO_2. If before drying the cellulose was coagulated in water, an additional step is needed. Water must be replaced by a fluid which is miscible with CO_2, such as acetone or ethanol. Finally, to obtain CAC, aerocellulose is pyrolysed [56–58]. The difference in the preparation of aerocellulose and classical aerogels is that no chemical cross-linking is made and the pore size distribution of aerocellulose is very wide: from a few nanometres to several microns.

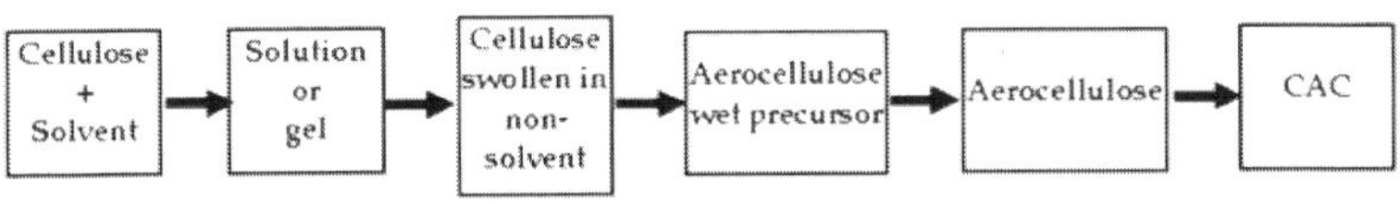

Figure 5.1 Main steps in the preparation of carbon aerocellulose.

Each of the preparation parameters (solvent, cellulose type, molecular weight and concentration, state of precursor mix (liquid or gel), type and temperature of the coagulation bath, pyrolysis conditions) influences the properties of both the aerocellulose and the final CAC, including their morphology and porosity [48]. This section focuses on the main parameters influencing the morphology and pore size of CAC and describes the results obtained for CAC prepared from microcrystalline cellulose dissolved in an 8 wt% NaOH solution, as it was found that these samples demonstrated the best electrochemical performance when used as supports for Pt electrocatalysts and also in primary lithium batteries.

An example of the macroscopic view of aerocellulose and its carbonised counterpart is shown in Fig. 5.2. The shape of the cylindrical mould remains the same in the pyrolysed sample despite a very high contraction, about 90%. It is thus possible to design monolithic CAs of any shape (determined by the shape of the mould), which is a distinct advantage of CAC over CAs based on cellulose acetate [42]. For CAs synthesised from cellulose acetate, the melting temperature is lower than the degradation temperature leading to the appearance of bubbles. It is possible to maintain a homothetic shape during pyrolysis, as reported for CAs based on resorcinol-formaldehyde; however, formaldehyde is not a green reagent and is also a known carcinogen.

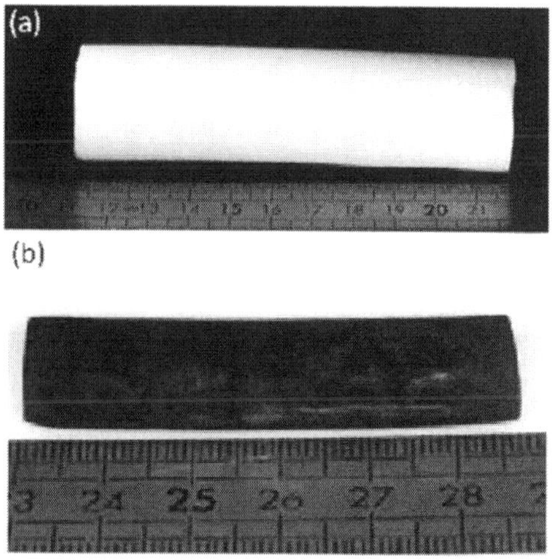

Figure 5.2 Cylindrical aerocellulose (a) and its pyrolysed counterpart, CAC (b).

The morphology of aerocellulose and the corresponding CAC is presented in Fig. 5.3. Aerocellulose has a net-like structure with the strands made of cellulose II. It has been shown that mechanically it behaves like an aerogel and not like a foam, with the Young modulus being proportional to the material density cubed [59]. The aerocellulose density varies from 0.1 to 0.3 g cm^{-3}, depending on the cellulose concentration in the initial solution. It is possible to make it lighter using a surfactant which creates air bubbles on mixing with the cellulose-NaOH solution [47], but these large pores collapse during pyrolysis. A large reduction in mass and volume occurs when aerocellulose is pyrolysed, causing densification: The density of CAC varies from 0.2 to 0.6 g cm^{-3} depending on cellulose concen-tration and pyrolysis conditions.

Pyrolysis conditions play an important role in CAC porosity and they also determine the shape of the sample. The maximum weight loss of cellulose during pyrolysis occurs at 300–350°C, as shown by the thermogravimetric analysis. During the pyrolysis of aerocellulose, the temperature should be increased slowly up to 330°C, then kept constant for a few hours and subsequently increased to the final reaction temperature (around 1000°C). Temperature stabilisation at 330°C is needed to allow the aerocellulose to slowly adapt (rearrangement of the carbon atoms) to the mass and volume loss. The latter induces strong internal stresses which may lead to the formation of cracks, sample deformation and even causing the sample to break up.

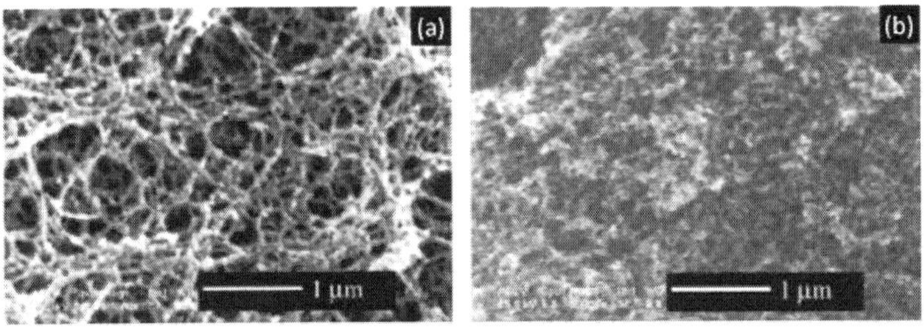

Figure 5.3 Aerocellulose (a) and CAC (b) derived from a 7 wt% cellulose in 8 wt% NaOH(aq) gel, coagulated in ethanol and dried under supercritical CO_2. The densities are 0.14 g cm^{-3} (a) and 0.3 g cm^{-3} (b). Reproduced by the permission of The Electrochemical Society from Ref. [57].

The porosity of CAC, its pore size distribution and the shape of its pores, govern its electrochemical performance. This porosity depends on the preparation conditions, with cellulose concentration playing one of the major roles. Unfortunately it is not possible to follow pore size evolution before and after pyrolysis, i.e. to compare the pore size distribution in aerocellulose and its carbon counterpart. The reason is that during mercury intrusion measurements aerocellulose is completely compressed under the pressure of mercury; nitrogen sorption techniques can only determine mesoporosity (<50 nm) down to microporosity, depending on the performance of the machine used. Thus the influence of cellulose concentration is shown only for CACs prepared from 5 wt% and 7 wt% microcrystalline cellulose dissolved in 8 wt% NaOH solution, gelled and coagulated in either a mixture of water:ethanol (1:1) or pure ethanol. Drying and pyrolysis conditions were the same for all samples.

Pore size distributions for the four samples mentioned above are shown in Fig. 5.4. For higher cellulose concentrations, smaller pore sizes were obtained, as expected. The key textural properties of CAC are given in Table 5.1. As will be shown later, the difference in pore size of CACs made from 5 wt% and 7 wt% cellulose determines the discharge properties of button cell primary lithium batteries and also plays a role in PEMFC performance owing to the influence a material's porosity has on diffusion properties of that material.

Table 5.1 Textural properties of CACs

Sample	Composition	Coagulation bath	Density (g cm^{-3})	Mesopore volume (cm^3 g^{-1})	Pore volume BJH (cm^3 g^{-1})	Micropore volume (cm^3 g^{-1})	BET specific surface (m^2 g^{-1})	Mean pore diameter (nm)
1	5 wt% cellulose – 8 wt% NaOH	water:ethanol = 1:1	0.20	4.36	0.69	0.063	340	83
2		ethanol	0.29	2.9	0.52	0.02	244	92
3	7 wt% cellulose – 8 wt% NaOH	water:ethanol = 1:1	0.26	3.17	0.52	0.017	297	61
4		ethanol	0.30	2.64	0.67	0.016	252	65
Saft reference (see details in Section 5.4)	from carbon black	—	0.30	2.6	0	0	42	70–80

Note: Sample numbers correspond to the ones indicated on the curves in Fig. 5.4.

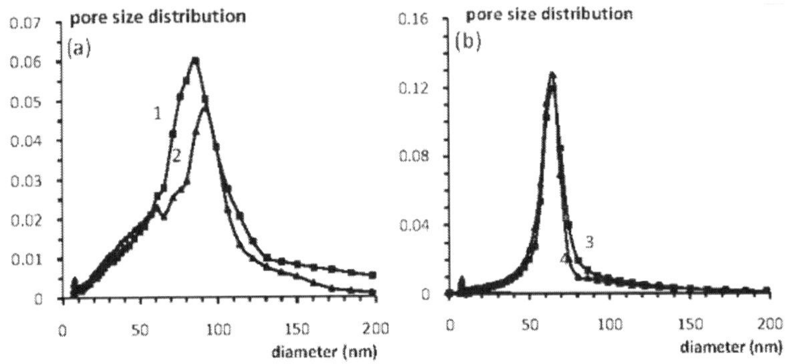

Figure 5.4 Pore size distribution of CACs prepared using gels obtained from 5 wt% (a) and 7 wt% (b) microcrystalline cellulose in 8 wt% NaOH(aq), coagulated in 1:1 water:ethanol (1, 3) or in ethanol alone (2, 4).

5.3 Utilisation of CAC as Pt Support for PEMFC Electrocatalysts

In proton exchange membrane fuel cells (PEMFC), the carbon substrate has multiple roles to achieve [64]. It needs to (i) provide a high dispersion of metal nanoparticles and protect them from agglomeration, (ii) be chemically inert and pure so that impurities do not inhibit electrochemical reactions, (iii) have high electron conductance to minimise ohmic losses associated with the electron transport, (iv) possess adequate porosity to ensure efficient mass transport of reactants to, and products from, the catalytic layers (CL), (v) be compatible with proton-conducting polymers so that composites containing a catalyst and a proton conductor provide efficient proton transport through the CL, (vi) provide optimum water management; (vii) be corrosion resistant and finally (viii) be obtained from an abundant source at minimal cost.

To be more specific, according to the previous list, the carbon substrate first consists of the inactive but electron-conductive support for the (usually Pt) electrocatalyst nanoparticles. The high cost and low availability of Pt means that a high electrochemical surface area (ECSA, expressed in $m^2\ g^{-1}{}_{Pt}$) is mandatory, which is only possible provided that a sufficient area is available for Pt nanoparticle anchoring. The carbon substrate should therefore display a large surface area (BET area) to enhance Pt dispersion and interact sufficiently with the Pt nanoparticles so as to favour their long-term dispersion. However, the substrate should not provoke any detrimental perturbation of the electronic properties of Pt, which could inhibit its (good) electrocatalytic properties, as TiO_x substrates have shown to do [65]. Second, its textural properties should be adapted to ionomer insertion (e.g. Nafion®) within the CL, whilst maintaining sufficient pore sizes of relevant geometry to enable both a high utilisation and high effectiveness factor for the Pt nanoparticles [19].

In practice, high-surface-area materials from the CB family are now the most widely used substrate material in PEMFC CLs and gas diffusing layers (GDL) [16, 17]. Despite their widespread (quasi-exclusive) use in PEMFC membrane electrode assemblies (MEA), they have two major drawbacks. From a performance point of view, CBs are not perfectly sizable materials, since their specific surface area calculated *via* BET theory is correlated to their grain size [66]: a larger specific surface area, compatible with high Pt dispersion, is coupled with a small mean pore size. This is incompatible with a high mass transport rate [21, 22]. In addition, their natural aggregation *via* van der Waals forces favours electronic insulation by encapsulation within the proton-conducting ionomer. Finally, these materials usually originate from the petroleum industry and can therefore not be considered green materials, as previously stated in the chapter.

Conversely, there are a wide range of high-surface-area carbon materials that could prove useful in PEMFC MEAs [64]. Amongst these, monolithic carbon structures fabricated from aerogels (CA) or xerogels (CX), are surely of interest. Owing to their unique structure, CAs display apt physical properties such as a high conductivity, high mesoporosity concomitant with a small degree of microporosity and a high surface area [28, 67, 68]. Of particular interest is the possibility to adjust the structure (tortuosity, pore volume, pore size distribution, specific surface area) and physical properties (density, conductivity) of the CA by varying controllable experimental parameters (e.g., precursor materials and their ratios, process parameters during sol–gel polymerisation and during pyrolysis), which has led to these materials being used with some success in PEMFC CLs [34, 69–71].

However, as for CB, CA obtained from resorcinol and formaldehyde (RF) is definitely not a green material. Consequently, changing the nature of the precursors used in CA synthesis, in particular using cellulose, an undoubtedly green biopolymer that is highly abundant on Earth, is very desirable. In fact cellulose has already been used with some success in PEMFC MEAs [72], notably owing to their reinforcing properties which are imparted on the polymer electrolyte membrane [73, 74]. Cellulose has also been investigated as a potential precursor for carbon materials used in the GDL [75, 76]. This section describes the use of CAC as a support for Pt nanoparticle deposition for use in PEMFC electrodes. CAC is certainly compatible with Pt nanoparticle deposition, owing to its high specific surface area (see Table 5.1), complete Nafion® intrusion, good O_2 mass transport and pore sizes in the large mesopore/small macropore range (pore diameter mainly in the 20–200 nm range as shown in Figs. 5.4a,b).

5.3.1 Pt Nanoparticle Deposition onto CAC Materials and Their Physicochemical/Electrochemical Characterisations

The CAC macromonolith obtained through pyrolysis was first broken down into smaller grains *via* manual grinding. The Pt nanoparticles were subsequently deposited on the CAC surface *via* an impregnation/reduction route. However, this route requires a control of the surface groups of the substrate, as these govern the interactions with the Pt ionic species. Unfortunately, CACs are not as clean a CA as those obtained from RF precursors. The early studies of Guilminot *et al.*, on cellulose acetate [55] or microcrystalline cellulose [56], have indeed demonstrated (by Fourier transform infrared spectroscopy [FTIR]) that numerous non-pure carbon groups (e.g. CH_3, CH_2, OH) were present after pyrolysis of the aerocellulose at ca. 800°C. To enable facile adsorption of the Pt ionic species (H_2PtCl_6 precursor was used, in a slightly acidic solution), two types of thermal treatments have been employed to favour the cleaning of the CAC surface: activation in a CO_2 atmosphere at 830°C [55, 56] or pyrolysis at 1050°C [57, 58], with both thermal treatments conducive to a higher carbon content on the surface of the CAC. Platinum nanoparticles were deposited onto the pre-ground CAC substrate (grains of a few micrometers which keep their initial texture) by impregnation with a predetermined amount of hexachloroplatinic acid,

$H_2PtCl_6 \cdot 6H_2O$, using the so-called strong electrostatic adsorption method [57, 58, 77]. In this procedure the pH of the acidic Pt^{z+}-containing solution was controlled to optimise the dispersion of Pt. The excess non-adsorbed hexachloroplatinic species were then washed away by filtration, because their presence favours Pt nanoparticle agglomeration (Fig. 5.5a), which is not suitable for PEMFC electrocatalysts [55, 56, 58]. Subsequently the adsorbed Pt^{z+} species /CAC were reduced to metal Pt nanoparticles/CAC, either by suspending the filtered and dried Pt^{z+}/CAC material in an alkaline $NaBH_4$ solution (Fig. 5.5b) or under flowing hydrogen at 300°C for 2 h (Fig. 5.5c). Afterwards the mixture was filtered and washed with hot water to remove any non-reduced platinum and boron containing species prior to drying overnight at 110°C. The latter technique yields very small Pt nanoparticles (Fig. 5.5c), with insufficient electrocatalytic properties [58], according to the well-known detrimental size effect of Pt nanoparticles with regard to the reduction of O_2 [78, 79]. On the contrary, the former favours nucleation of Pt nanoparticles with diameters of ca. 3–4 nm (Fig. 5.5b) [57, 58], the size of which is suitable for electrocatalysis [80, 81].

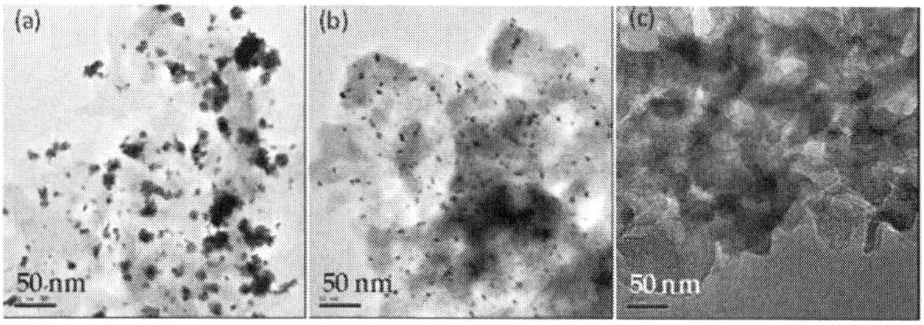

Figure 5.5 TEM images showing the morphology of the Pt/CAC nanoparticles after impregnation by H_2PtCl_6 and subsequent reduction by $NaBH_4$ without filtration (a) or in $NaBH_4$ (b) or H_2 at 300°C (c), both after filtration. Reproduced from Ref. [58] with permission from The Electrochemical Society.

Nevertheless, in all cases, the amount of Pt immobilised onto the CAC surface was insufficient (<10 wt% Pt/(Pt + C)), and one final sample was prepared by multiple impregnation/reduction steps for fuel cell testing experiments. The best as-prepared sample, containing ca. 14 wt% Pt/(Pt + C) was obtained in three successive impregnation/reduction steps, the first reduction being achieved in an alkaline $NaBH_4$ solution to favour nucleation of Pt nanoparticles of a suitable diameter (ca. 3 nm) and the subsequent steps under H_2 at 400°C to prevent their agglomeration and maintain adequate Pt dispersion. This sample has been thoroughly characterised by ICP-AES, XRD and TEM, as highlighted in Fig. 5.6 [58].

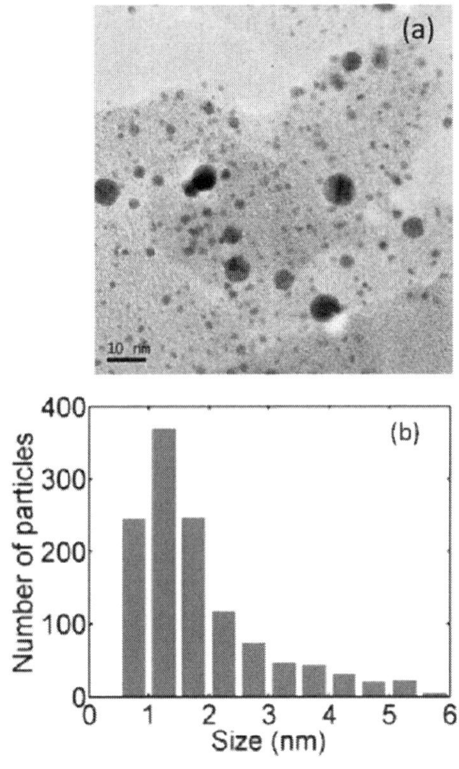

Figure 5.6 TEM images showing the morphology of the Pt/CAC nanoparticles (a) and corresponding particle size distribution histograms (b) after three impregnation steps with H_2PtCl_6, filtering and subsequent reduction with $NaBH_4$ (first impregnation) or H_2 at 400°C (second and third impregnations). Reproduced from Ref. [58] with permission from The Electrochemical Society.

The electrocatalyst obtained by this technique, labelled Pt/CAC, possesses sufficient ECSA (77 $m^2\ g^{-1}{}_{Pt}$ obtained from CO-stripping coulometry versus 68 $m^2\ g^{-1}{}_{Pt}$ from TEM imaging—the closeness of the values for the ECSA obtained by CO-stripping coulometry and TEM imaging demonstrate the homogeneity of the Pt/CAC sample) and electrocatalytic properties to be tested in a PEMFC unit cell. This sample is comparable to state-of-the-art Pt/CB electrocatalysts from Tanaka TKK in terms of electrocatalytic activity (Table 5.2).

Table 5.2 Electrochemical surface area of Pt (ECSA/$m^2\ g^{-1}{}_{Pt}$, as obtained from CO-stripping coulometry) and oxygen reduction reaction (ORR) kinetic parameters (specific activity, SA/$\mu A\ cm^{-2}{}_{Pt}$ or mass activity, MA/$A\ g^{-1}{}_{Pt}$) for the Pt/CAC samples synthesised by three successive impregnation/reduction steps

Sample	ECSA	MA -0.85 V	MA -0.9 V	SA -0.85 V	SA -0.9 V
Pt/CAC, 14 wt%	77	36	8.5	47	11
Pt/CB, 67 wt% (TKK)	62	32	5.6	52	9.1

Note: The result for a commercial Pt/CB material from Tanaka TKK is given for comparison.

5.3.2 PEMFC Unit Cell Testing of the Pt/CAC Materials

PEMFC polarisation data were acquired on a Biologic automated test bench (FCT-150Z, Bio-Logic). In all the tests, dry H_2/humidified air were fed to the anode and cathode, respectively. The gas pressures were kept identical at the two electrodes (P = 2/2 bars relative, i.e. 3/3 bars absolute pressure) and the stoichiometry was set at $S_{H_2} = 1.1/s_{air} = 2$. In all cases, the membrane electrode assemblies were prepared at Paxitech using a proprietary procedure, with a commercial Nafion®-like proton exchange membrane and commercial gas diffusion layers. The anode was based on commercial Pt/CB (67 wt% from Tanaka Kikinzoku Kogyo, TKK) at a Pt loading of 0.5 $mg_{Pt}\ cm^{-2}$, while the cathode contained the Pt/CAC obtained from three successive impregnation reduction steps or a commercial Pt/CB. More details about these procedures can be found in Ref. [58].

The appropriate Nafion® to carbon ratio was first determined for the Pt/CAC material. The best effectiveness factor (ε_{Pt}) [19, 82] is obtained for N/C = 1.1, although Pt utilisation factor (u_{Pt}) is larger for N/C = 0.77 [24].

The unit cell tests show that the Pt/CAC material can indeed be used with some success in PEMFCs. This was, to the authors' knowledge, the first practical demonstration. However, whereas, Pt/CAC competes favourably with Pt/CB when the Pt loadings are comparable (14 versus 8.8 wt% Pt/(Pt + C) for Pt/CAC and Pt/CB), the Pt/CAC is non-negligibly outperformed by highly loaded Pt/CB (67 wt% by Tanaka TKK) (Fig. 5.7). The rationale for the better overall performances of Pt/CAC versus Pt/CB (8.8 wt%) is likely to originate from the better porous structure of the Pt/CAC; the intrinsic activity, shown at current densities below 0.4 A cm^{-2}, is indeed similar for these two materials, while the mass transport limitations (measured at high current density, i.e. above 0.4 A cm^{-2}) are non-negligibly larger for Pt/CB, thereby ruining its activity at high load. On the contrary, Pt/CB 67 wt% MEA are not so detrimentally inhibited by mass transport, because the high Pt/(Pt + C) loading of this material is compatible with the very small thickness of the CL (for all cathode CL, the Pt content is fixed at ca. 0.5 mg_{Pt} cm^{-2}, the anode, membrane and GDL being kept similar). By comparison, the Pt/CAC (14 wt% MEA) displays a ca. 5 times thicker CL within which the mass transport limitations are more pronounced.

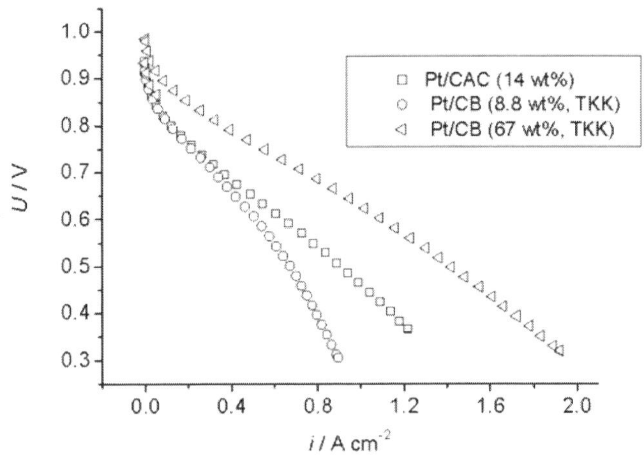

Figure 5.7 PEMFC unit cell testing for the Pt/CAC (4 wt%) cathode, as compared with Pt/CB (67 wt% TKK) or Pt/CB (8.8 wt% TKK) cathodes. In all cases, the Pt load at the cathode was ca. 0.5 mg_{Pt} cm^{-2} and the anode uses a standard Paxitech electrode with Pt/CB 67 wt% TKK electrocatalyst. P_{air}/P_{H_2} = 2.0/2.0 bar (relative); s_{air}/s_{H_2} = 2.0/1.1, d_{minair}/d_{minH_2} = 300/120 mL min^{-1}. T = 70°C (air RH50%).

While the feasibility of using such materials in PEMFC electrodes has been clearly demonstrated, one improvement would be to maximise the Pt/(Pt + C) loading over the CAC substrate, to values exceeding 30–40 wt%.

5.4 Utilisation of CAC as Cathode Material in Li/SOCl$_2$ Primary Batteries

Carbons have historically been used as electrically conducting additives in batteries [83] from the old Zn/MnO_2 primary batteries to more recent developments such as carbon enhanced valve regulated lead acid (VRLA) secondary batteries [84] or positive electrodes for Li ion batteries. Carbons are also the main component of positive electrodes in metal/air batteries [85, 86] and of both electrodes in organic supercapacitors [87], where very high surface areas and tailored porosities are the key parameters. Carbons constitute the active material itself in the negative electrode in lithium ion batteries due to their ability to intercalate or absorb lithium ions at low electrochemical potential [88–90]. Carbons are the main component of the positive electrode in primary lithium/ liquid cathode batteries as well [91].

The latter family of primary batteries presents outstanding characteristics, the main one being that the reducer (lithium) is in direct contact with the liquid oxidant (thionyl chloride $SOCl_2$, sulphuryl chloride SO_2Cl_2 or sulphur dioxide SO_2 dissolved in acetonitrile). Contrary to classical batteries, the electrolyte is also electroactive, so the amount of inert material is reduced. Combined with their high voltage (3.5 V for Li/$SOCl_2$), it results in the highest energy density for primary batteries, up to 600 Wh kg^{-1} for Li/$SOCl_2$, compared to 150 Wh kg^{-1} for the best alkaline primary batteries. The presence of a strongly oxidising and corrosive liquid limits the use of such batteries in the consumer market, but they find extensive applications in military and industrial markets. Automatic toll gates, reading electricity meters, alarms, pressure sensors, GPS, emergency exit signs, night vision and communication devices are examples of their current applications. The overall market for Lithium/liquid cathode batteries is far from negligible and in 2007 the production of Li/$SOCl_2$ cells by Saft (France) was ca. 70 million.

The viability of such batteries relies on the insolubility of one of the oxido-reduction products (LiCl in the case of thionyl and sulphuryl chloride). This insoluble product precipitates immediately on the surface of the lithium electrode when filling the cell during fabrication and prevents further direct contact between electroactive species [91]. The second singularity is that the current collector, although not being electrochemically active by itself, plays a key role in the performance of the battery. In solid cathode cells the current collection is made of a simple inert metal grid or foil, with minimal contribution to electrode weight or volume. In contrast, in liquid cathodes systems, the current collector must also accommodate the insoluble reaction products within its porous network yet still ensure a high surface area that is in contact with the liquid throughout the discharge. Thus the current collector must present both a high surface area and a high porosity in the range of 70–90%. It should also be corrosion resistant to highly oxidising liquids.

The importance of the current collector can be seen in Fig. 5.8, where a $Li/SOCl_2$ cylindrical cell with concentric electrodes (bobbin cell) is schematically represented. The carbon block containing the liquid cathode is 85% porous and occupies 70% of the inner volume of the cell. Although helix-type cells also exist, bobbin-type cells are by far the major type of $Li/SOCl_2$ cells.

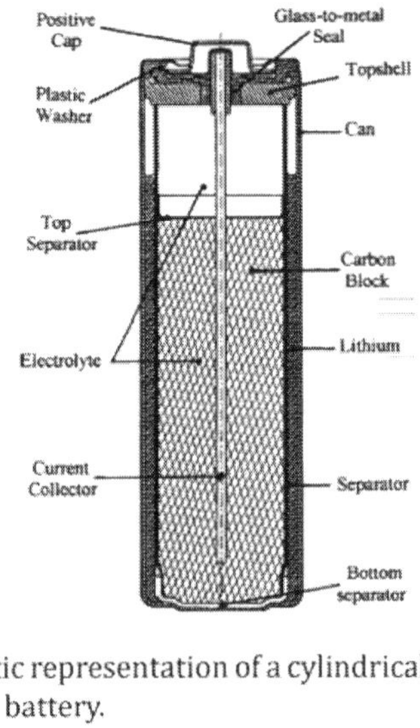

Figure 5.8 Schematic representation of a cylindrical bobbin-type $Li/SOCl_2$ primary battery.

5.4.1 Interest of using CAC as Cathode Material in Li/SOCl$_2$ Primary Batteries

The carbon block of the cell with suitable surface area and porosity is usually made from CB powders with a polytetrafluoroethylene (PTFE) polymeric binder. The nature of the carbon and its influence on the resultant physical characteristics of the electrode have been extensively studied [92–96]. In contrast to classical batteries that have a solid cathode and anode where the discharge is limited by either the amount of oxidant or reductant, in liquid cathode cells it can be limited by the carbon collector whenever the carbon porosity is completely blocked by the insoluble product. Thus, the cell capacity is primarily dependent on the carbon collector porosity [94].

A precise analysis of the influence of pore size upon LiCl precipitation has been carried out using different techniques [97–99]. It seems that the workable range of pore diameters that can accommodate LiCl is quite wide, from large micropores (1–2 nm) through to mesopores (2–50 nm) and to small macropores (<100 nm). Thus, the optimum carbon current collector will present the highest possible porous volume in this pore size range.

CA monoliths can be realised with characteristics as described above [25, 29] and are thus suitable candidates as current collectors for Li/SOCl$_2$ primary batteries. There are many possible advantages for using CA cylinders instead of the current CB-based electrode, the main ones being cost, environmental considerations and electrochemical performances. It has been shown by Saft [100] that CAs from resorcinol-formaldehyde precursors could be synthesised as cylindrical monoliths with a porosity of over 90%. It has been demonstrated that they could replace and even exceed, in terms of performances, the existing carbon electrode. However, the cost objective forecasted for this family of CAs has inhibited further investigation into their use in complete cells.

Today the implementation of the carbon electrode consists of a relatively complex industrial process of mixing CB powder (from thermal treatment of compounds from the oil industry) with a polymeric binder dispersion, followed by extrusion or compression and eventually solvent drying. The resulting cylinder must present sufficient mechanical properties to be manipulated by automated machines during cell assembly. Thus, one can easily conceive that the simplification due to the use of ready-made CAC cylindrical monoliths could lead to cost reductions in production, even if the cost of raw CB is low. The final cost of the classical CB electrode, taking into account all of the fabrication steps, has been estimated at around \$10 kg^{-1}. CAs from cellulose may theoretically match this cost since cellulose extraction and production of dissolving pulps is a well-known and established industrial process. It is obvious that from an environmental point of view the use of cellulose has a lot of advantages over CB. As shown in Table 5.1, the main contribution to the monolith's porosity stems from the small macropores; however, CACs also have micropores and mesopores, contrary to the Saft CB reference electrode. Some CAC samples present significantly lower densities than the Saft reference, which results in a better electrochemical performance, as outlined in the following section.

5.4.2 Electrochemical Characteristics of CAC in Li/$SOCl_2$ Primary Batteries

The first set of electrical characterisations of CAC monoliths have been conducted on thin (<1 mm) slices of cylinders in button-type cells (Fig. 5.9). This cell is conceived with both an electrolyte and a lithium excess, which limits discharge by blocking the pores of the carbon current collector.

Figure 5.9 Button-type welded cell fabricated with CAC for electrochemical testing.

The discharge curves (Figs. 5.10a,b) confirm that CACs are suitable current collectors and in certain cases their capacity/volume was higher than the reference (Fig. 5.10a), which was expected owing to their lower densities. It is interesting to note that the performance of a CAC made from a 5 wt% cellulose solution (Fig. 5.10a) is better than the CAC prepared using 7 wt% cellulose (Fig. 5.10b). The reason for this is the difference in the porosity of the samples. Despite the fact the mesopore volume is practically the same, the size of pores in CAC prepared using 5 wt% cellulose is higher than the pore size of CAC prepared using 7 wt% cellulose, 80–90 nm versus 60 nm, respectively (see Table 5.1 and Fig. 5.4). The results obtained show that by modifying the initial formulation of the CAC precursor it is possible to control the electrochemical properties of a current collector in a $Li/SOCl_2$ primary battery.

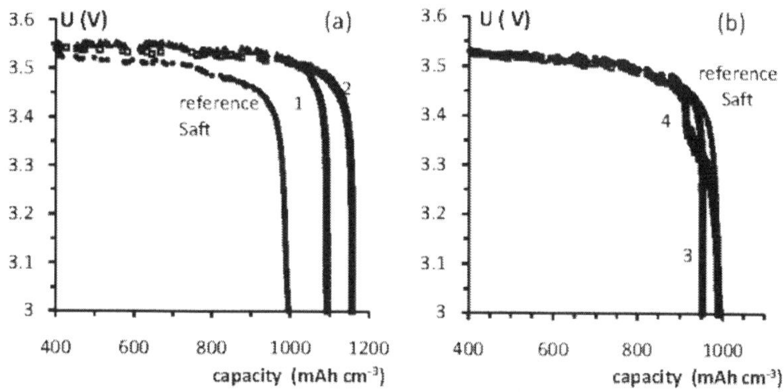

Figure 5.10 Discharge at 0.4 mA of $Li/SOCl_2$ button-type cells with a thin CAC collector disk: CAC samples 1, 2 (a) and 3, 4 (b) as described in Table 5.1 and Fig. 5.4. The discharge time was in the range of 300–400 h.

Second, the suitability of CAC as a current collector was also examined by implementing large CAC monoliths in industrial cylindrical 1/2 AA cells (14 mm diameter and 50 mm height). The discharge results were successful: the capacity of CAC was similar to the Saft reference electrodes and in some cases even higher. However, the advantage of CAC material in AA cells was less evident than in button-type batteries, as some technical problems still need to be solved. There is a build-up of internal pressure inside the carbon collector as the LiCl precipitates within the porous network. Classical CB powder-based electrodes swell up to 30% during LiCl precipitation. CAC cylinders are rigid and their mechanical integrity throughout discharge may sometimes be questioned. The swelling of classical CB electrodes compensates for the consumption of the lithium electrode during discharge, thus maintaining mechanical contact between the constituent parts of the cell.

Another technical problem was the electrical connection of the CAC to the external positive pole. In the case of plastic CB electrodes, the problem is simply solved by thrusting a metal pin into the CB material (Fig. 5.8), which is impossible for CAC cylinders. In order to keep good electrical contact, a small hole (1 mm diameter) was drilled in the CAC, filled with a plastic CB electrode and then the pin was inserted into this hole. This insured electrical contact but is obviously not a solution on an industrial scale.

In summary, the possibility of replacing CB electrodes by CAC monoliths in Li/liquid cathode cells have been assessed, and the following advantages can be listed:

(i) possibility of quasi directly preparing electrodes with a given shape and dimension (cylindrical, monolithic) by the pyrolysis of organic monoliths
(ii) easy assembly of the battery
(iii) no impurities (by controlling reagent quality)
(iv) porosity tailored to the application
(v) high intrinsic conductivity (no binding agent and covalent links between carbon particles)
(vi) good mechanical properties
(vii) lower cost and green reagents

Although this is promising, a thorough cost analysis still remains to be done, and some technical problems need to be resolved prior to industrial applications.

5.5 Conclusions

This work shows the possibilities and advantages of the use of cellulose-based CAs for electrochemical applications. Carbon monoliths were prepared from cellulose-NaOH(aq) gels that were coagulated, dried under supercritical CO_2 conditions and pyrolysed. Sample porosity and shape were controlled both by cellulose concentration and the parameters of each preparation step. Two applications were considered: proton exchange membrane fuel cells (PEMFC) and primary batteries ($Li/SOCl_2$). In the first case,

Pt nanoparticles can be homogeneously deposited within the CAC; successive impregnation and reduction steps are needed to optimise their size, distribution and loading. The tailored texture yields a porosity which reduces mass transport losses within the electrode structure. The new electrocatalyst may enable the amount of Pt in PEMFC electrodes to be minimised whilst increasing cell performance; the results are promising when compared with classical Pt/CB materials. In the second application, cellulose-based carbon monoliths were also successfully tested in button and industrial cylindrical 1/2 AA cells. The discharge results showed that these new carbons are suitable to use as cathode electrodes and a tailored texture leads to a higher capacity/volume than the reference matrix. The advantage of cellulose-based CAs is that they are based on a renewable resource (cellulose); their shape can be predetermined by the gel mould; formulation and synthesis routes allow control over the carbon porosity and importantly they have very promising electrochemical performances.

References

1. Che, G. L., Lakshmi, B. B., Fisher, E. R., and Martin, C. R. (1998). Carbon nanotubule membranes for electrochemical energy storage and production, *Nature*, **393**, pp. 346–349.

2. Joo, S. H., Choi, S. J., Oh, I., Kwak, J., Liu, Z., Terasaki, O., and Ryoo, R. (2001). Ordered nanoporous arrays of carbon supporting high dispersions of platinum nanoparticles, *Nature*, **412**, pp. 169–172.

3. Arico, A. S., Bruce, P., Scrosati, B., Tarascon, J. M., and Van Schalkwijk, W. (2005). Nanostructured materials for advanced energy conversion and storage devices, *Nat. Mater.*, **4**, pp. 366–377.

4. Frackowiak, E., and Beguin, F. (2001). Carbon materials for the electrochemical storage of energy in capacitors, *Carbon*, **39**, pp. 937–950.

5. Pozio, A., De Francesco, M., Cemmi, A., Cardellini, F., and Giorgi, L. (2002). Comparison of high surface Pt/C catalysts by cyclic voltammetry, *J. Power Sources*, **105**, pp. 13–19.

6. Auer, E., Freund, A., Pietsch, J., and Tacke, T. (1998). Carbons as supports for industrial precious metal catalysts, *Appl. Catal. A-Gen.*, **173**, pp. 259–271.

7. Boehm, H. P. (1994). Some aspects of the surface-chemistry of carbon-blacks and other carbons, *Carbon*, **32**, pp. 759–769.

8. Rodríguez-Reinoso, F., and Molina-Sabio, M. (1992). Activated carbons from lignocellulosic materials by chemical and/or physical activation: An overview, *Carbon*, **30**, pp. 1111–1118.

9. Baughman, R. H., Zakhidov, A. A., and de Heer, W. A. (2002). Carbon nanotubes-the route toward applications, *Science*, **297**, pp. 787–792.

10. Kroto, H. W. (1987). The stability of the fullerenes C-24, C-28, C-32, C-36, C-50, C-60 and C-70, *Nature*, **329**, pp. 529–531.

11. Ryoo, R., Joo, S. H., Kruk, M., and Jaroniec, M. (2001). Ordered mesoporous carbons, *Adv. Mater.*, **13**, pp. 677–681.

12. Pekala, R. W., Alviso, C. T., Kong, F. M., and Hulsey, S. S. (1992). Aerogels derived from multifunctional organic monomers, *J. Non-Cryst. Solids*, **145**, pp. 90–98.

13. Lin, C., and Ritter, J. A. (1997). Effect of synthesis pH on the structure of carbon xerogels, *Carbon*, **35**, pp. 1271–1278.

14. Maillard, F., Simonov, P. A., and Savinova, E. R. (2009). Carbon materials as supports for fuel cell electrocatalysts, in *Carbon Materials for Catalysis* (Serp, P., and Figueiredo, J. L., eds.), John Wiley & Sons, New York, Chapter 12, pp. 429–481.

15. Xia, K., Gao, Q., Jiang, J., and Hu, J. (2008). Hierarchical porous carbons with controlled micropores and mesopores for supercapacitor electrode materials, *Carbon*, **46**, pp. 1718–1726.

16. Vielstich, W., Lamm, A., and Gasteiger, H. A. (eds.) (2003). *Handbook of Fuel Cells*, Wiley, Chichester, Vol. 1–4, pp. 1–1372.

17. Gasteiger, H. A., Vielstich, W., and Yokokawa, H. (eds.) (2009). *Handbook of Fuel Cells*, John Wiley & Sons Ltd, Chichester, Vol. 5–6, pp. 1–1038.

18. United States Department of Energy Report (2010). *Critical Materials Strategy* (http://www.energy.gov/news/documents/criticalmaterials strategy.pdf).

19. Chatenet, M., Dubau, L., Job, N., and Maillard F. (2010). The (electro) catalyst|membrane interface in the proton exchange membrane fuel cell: Similarities and differences with non-electrochemical catalytic membrane reactors, *Catal. Today*, **156**, pp. 76–86.

20. Uchida, M., Aoyama, Y., Eda, N., and Ohta, A. (1995). Investigation of the microstructure in the catalyst layer and effects of both perfluorosulfonate ionomer and PTFE-loaded carbon on the catalyst layer of the polymer electrolyte fuel cells, *J. Electrochem. Soc.*, **142**, pp. 4143–4149.

21. Uchida, M., Aoyama, Y., Tanabe, M., Yanagihara, N., Eda, N., and Ohta, A. (1995). Influences of both carbon supports and heat-treatment of supported catalyst on electrochemical oxidation of methanol, *J. Electrochem. Soc.*, **142**, pp. 2572–2576.

22. Uchida, M., Fukuoka, Y., Sugawara, Y., Eda, N., and Ohta, A. (1996). Effects of microstructure of carbon support in the catalyst layer on the performance of polymer-electrolyte fuel cells, *J. Electrochem. Soc.*, **143**, pp. 2245–2252.

23. Gavalda, S., Gubbins, K. E., Hanzawa, Y., Kaneko, K., and Thomson, K. T. (2002). Nitrogen adsorption in carbon aerogels: A molecular simulation study, *Langmuir*, **18**, pp. 2141–2151.

24. Gavalda, S., Kaneko, K., Thomson, K. T., and Gubbins, K. E. (2001). Molecular modeling of carbon aerogels, *Colloids Surf. A*, **187–188**, pp. 531–538.

25. Pekala, R. W. (1989). Organic aerogels from the polycondensation of resorcinol with formaldehyde, *J. Mater. Sci.*, **24**, pp. 3221–3227.

26. Lu, X., Nilsson, O., Fricke, J., and Pekala, R. (1993). Thermal and electrical conductivity of monolith carbon aerogels, *J. Appl. Phys.*, **73**, pp. 581–584.

27. Hanzawa, Y., Kaneko, K., Yoshizawa, N., Pekala, R. W., and Dresselhaus, M. S. (1998). The pore structure determination of carbon aerogels, *Adsorption*, **4**, pp. 187–196.

28. Tamon, H., Ishizaka, H., Mikami, M., and Okazaki, M. (1997). Porous structure of organic and carbon aerogels synthesized by sol-gel polycondensation of resorcinol with formaldehyde, *Carbon*, **35**, pp. 791–796.

29. Pekala, R. W., and Kong, F. M. (1989). Resorcinol-formaldehyde aerogels and their carbonized derivatives, *Abstr. Pap. Am. Chem. Soc.*, **197**, pp. 113-Poly.

30. Job, N., Berthon-Fabry, S., Chatenet, M., Marie, J., Brigaudet, M., and Pirard, J.-P. (2009). Nanostructured carbons as platinum catalyst supports for proton exchange membrane fuel cell electrodes, *Top. Catal.*, **52**, pp. 2117–2122.

31. Job, N., Lambert, S., Chatenet, M., Gommes, C. J., Maillard, F., Berthon-Fabry, S., Regalbuto, J. R., and Pirard, J.-P. (2010). Preparation of highly loaded Pt/carbon xerogel catalysts for proton exchange membrane fuel cells by the strong electrostatic adsorption method, *Catal. Today*, **150**, pp. 119–127.

32. Job, N., Maillard, F., Marie, J., Berthon-Fabry, S., Pirard, J.-P., and Chatenet, M. (2009). Electrochemical characterization of Pt/carbon xerogel and Pt/carbon aerogel catalysts: First insights into the influence of the carbon texture on the Pt nanoparticle morphology and catalytic activity, *J. Mater. Sci.*, **44**, pp. 6591–6600.

33. Marie, J., Berthon-Fabry, S., Achard, P., Chatenet, M., Chainet, E., Pirard, R., and Cornet, N. (2006). Synthesis of highly porous catalytic layers for polymer electrolyte fuel cell based on carbon aerogels, *ECS Trans.*, **1**, pp. 509–519.

34. Marie, J., Berthon-Fabry, S., Achard, P., Chatenet, M., Pradourat, A., and Chainet, E. (2004). Highly dispersed platinum on carbon aerogels as supported catalysts for PEM fuel cell-electrodes: Comparison of two different synthesis paths. *J. Non-Cryst. Solids*, **350**, pp. 88–96.

35. Marie, J., Berthon-Fabry, S., Chatenet, M., Chainet, E., Pirard, R., Cornet, N., and Achard, P. (2007). Platinum supported on resorcinol-formaldehyde based carbon aerogels for PEMFC electrodes: Influence of the carbon support on electrocatalytic properties, *J. Appl. Electrochem.*, **37**, pp. 147–153.

36. Marie, J., Chenitz, R., Chatenet, M., Berthon-Fabry, S., Cornet, N., and Achard, P. (2009). Highly porous PEM fuel cell cathodes based on low density carbon aerogels as Pt-support: Experimental study of the mass-transport losses, *J. Power Sources*, **190**, pp. 423–434.

37. Smirnova, A., Dong, X., Hara, H., and Sammes, N. (2006). Aerogel-based PEMFC catalysts operating at room temperature, *J. Fuel Cell Sci. Tech.*, **3**, pp. 477–481.

38. Smirnova, A., Dong, X., Hara, H., Vasiliev, A., and Sammes, N. (2005). Novel carbon aerogel-supported catalysts for PEM fuel cell application, *Int. J. Hydrogen Energy*, **30**, pp. 149–158.

39. Smirnova, A., Wender, T., Goberman, D., Hu, Y. L., Aindow, M., Rhine, W., and Sammes, N. M. (2009). Modification of carbon aerogel supports for PEMFC catalysts, *Int. J. Hydrogen Energy*, **34**, pp. 8992–8997.

40. Tan, C., Fung, B. M., Newman, J. K., and Vu, C. (2001). Organic aerogels with very high impact strength, *Adv. Mater.*, **13**, pp. 644–646.

41. Fischer, F., Rigacci, A., Pirard, R., Berthon-Fabry, S., and Achard, P. (2006). Cellulose-based aerogels, *Polymer*, **47**, pp. 7636–7645.

42. Hildenbrand, C. (2010). PhD Thesis, MinesParisTech, Nanostructured carbons from cellulose-derivative based aerogels for electrochemical energy storage and conversion: Evaluation as EDLC electrode material.

43. Grzyb, B., Hildenbrand C., Berthon-Fabry, S., Bégin, D., Job, N., Rigacci, A., and Achard, P. (2010). Functionalisation and chemical characterisation of cellulose derived carbon aerogels, *Carbon*, **48**, pp. 2297–2307.

44. Salam, A., Venditti, R. A., Pawlak, J. J., and El-Tahlawy, K. (2011). Crosslinked hemicellulose citrate–chitosan aerogel foams, *Carbohydr. Polym.*, **84**, pp. 1221–1229.

45. Szczurek, A., Amaral-Labat, G., Fierro, V., Pizzi, A., Masson, E., and Celzard, A. (2011). The use of tannin to prepare carbon gels. Part I: Carbon aerogels, *Carbon*, **49**, pp. 277–278.

46. Innerlohinger, J., Weber, H. K., and Kraft, G. (2006). Aerocellulose: Aerogel and aerogel like materials made from cellulose, *Macromol. Symp.*, **244**, pp. 126–135.

47. Gavillon, R., and Budtova, T. (2008). Aerocellulose: New highly porous cellulose prepared from cellulose-NaOH aqueous solutions, *Biomacromolecules*, **9**, pp. 269–277.

48. Sescousse, R. (2010). PhD Thesis, MinesParisTech, Nouveaux matériaux cellulosiques ultra poreux et leurs carbones à partir de solvants verts. (in French).

49. Cotorueloa, L. M., Marquésa, M. D., Rodríguez-Mirasola, J., Rodríguez, J. J., and Cordero, T. (2009). Lignin-based activated carbons for adsorption of sodium dodecylbenzene sulfonate: Equilibrium and kinetic studies, *J. Colloid Interface Sci.*, **332**, 1, 39–45.

50. Jin, H., Nishiyama, Y., Wada, M., and Kuga, S., (2004). Nanofibrillar cellulose aerogels, *Colloid Surf. A*, **240**, pp. 63–67.

51. Aaltonen, O., and Jauhiainen, O. (2009). The preparation of lignocellulosic aerogels from ionic liquid solutions, *Carbohydr. Polym.*, **75**, pp. 25–129.

52. Valentin, R., Molvinger, K., Quignard, F., and Di Renzo F. (2005). Methods to analyse the texture of alginate aerogel microspheres? *Macromolecul. Symp.*, **222**, pp. 93–102.

53. Glenn, G. M., and Irving, D. W. (1995). Starch-based microcellular foams, *Cereal Chem.*, **72**, pp. 155–161.

54. Tsioptsias, C., Michailof, C., Stauropoulos, G., and Panayiotou, C. (2009). Chitin and carbon aerogels from chitin alcogels, *Carbohydr. Polym.*, **76**, pp. 535–540.

55. Guilminot, E., Fischer, F., Chatenet, M., Rigacci, A., Berthon-Fabry, S., Achard, P., and Chainet, E. (2007). Use of cellulose-based carbon aerogels as catalyst support for PEM fuel cell electrodes: Electrochemical characterisation, *J. Power Sources*, **166**, pp. 104–111.

56. Guilminot, E., Gavillon, R., Chatenet, M., Berthon-Fabry, S., Rigacci, A., and Budtova, T. (2008). New nanostructured carbons from porous cellulose: Elaboration, pyrolysis and subsequent use as substrate for proton exchange membrane fuel cell electrocatalyst, *J. Power Sources*, **185**, pp. 717–726.

57. Rooke, J., Matos, C., Chatenet, M., Sescousse, R., Budtova, T., Berthon-Fabry, S., Mosdale, R., and Maillard, F. (2010). Elaboration and characterizations of platinum nanoparticles supported on cellulose-based carbon aerogel, *ECS Trans.*, **33**, pp. 447–459.

58. Rooke, J., Matos, C., Chatenet, M., Sescousse, R., Budtova, T., Berthon-Fabry, S., Mosdale, R., and Maillard, F. (2011). Evaluation of platinum nanoparticles supported on cellulose-based carbon aerogel in PEMFC unit cells, *J. Electrochem. Soc.*, **158**, pp. B779–B789.

59. Sescousse, R., Gavillon, R., and Budtova, T. (2011). Aerocellulose from cellulose-ionic liquid solutions: Preparation, properties and comparison with cellulose-NaOH and cellulose-NMMO routes, *Carbohydr. Polym.*, **83**, pp. 1766–1774.

60. Sescousse, R., and Budtova, T. (2009). Influence of processing parameters on regeneration kinetics and morphology of porous cellulose from cellulose-NaOH-water solutions, *Cellulose*, **16**, pp. 417–426.

61. Sescousse, R., Smacchia, A., and Budtova, T. (2010). Influence of lignin on cellulose-NaOH-water mixtures properties and on Aerocellulose morphology, *Cellulose*, **17**, pp. 1137–1146.

62. Sescousse, R., Gavillon, R., and Budtova T. (2011). Wet and dry highly porous cellulose beads from cellulose-NaOH-water solutions: Influence of the preparation conditions on beads shape and encapsulation of inorganic particles, *J. Mater. Sci.*, **46**, pp. 759–765.

63. Roy, C., Budtova, T., and Navard, P. (2003). Rheological properties and gelation of aqueous cellulose-NaOH solutions, *Biomacromolecules*, **4**, pp. 259–264.

64. Maillard, F., Simonov, P., and Savinova, E. R. (2009). Carbon materials as support for fuel cells electrocatalysts, in *Carbon Materials for Catalysis* (Serp, P., and Figueiredo, J. L., eds.), John Wiley & Sons, Inc., New York, Vol. 5, pp. 429–480.

65. Micoud, F. (2009). PhD Thesis, Grenoble-INP, Influence d'un support MO_x (M = W, Ti) sur les propriétés électrocatalytiques de nano-particules de platine (in French).

66. Uchida, M., Fukuoka, Y., Sugawara, Y., Ohara, H., and Ohta, A. (1998). Improved preparation process of very-low-platinum-loading electrodes for polymer electrolyte fuel cells, *J. Electrochem. Soc.*, **145**, pp. 3708–3713.

67. Pekala, R. W., Farmer, J. C., Alviso, C. T., Tran, T. D., Mayer, S. T., Miller J. M., and Dunn, B. (1998). Carbon aerogels for electrochemical applications, *J. Non-Cryst. Solids*, **225**, pp. 74–80.

68. Reichenauer, G., Emmerling, A., Fricke, J., and Pekala, R. W. (1998). Microporosity in carbon aerogels, *J. Non-Cryst. Solids*, **225**, pp. 210–214.

69. Glora, M., Wiener, M., Petricevic, R., Probstle, H., and Fricke, J. (2001). Integration of carbon aerogels in PEM fuel cells, *J. Non-Cryst. Solids*, **285**, pp. 283–287.

70. Petricevic, R., Glora, M., and Fricke, J. (2001). Planar fibre reinforced carbon aerogels for application in PEM fuel cells, *Carbon*, **39**, pp. 857–867.

71. Marie, J., Berthon-Fabry, S., Achard, P., Chatenet, M., Chainet, E., Pirard, R., and Cornet, N. (2006). Synthesis of highly porous catalytic layers for polymer electrolyte fuel cell based on carbon aerogels, *ECS Trans.*, **1**, pp. 509–519.

72. Shuhaimi, N. E. A., Alias, N. A., Kufian, M. Z., Majid, S. R., and Arof, A. K. (2010). Characteristics of methyl cellulose-NH_4NO_3-PEG electrolyte and application in fuel cells, *J. Solid State Electrochem.*, **14**, pp. 2153–2159.

73. Samir, M., Alloin, F., and Dufresne, A. (2005). Review of recent research into cellulosic whiskers, their properties and their application in nanocomposite field, *Biomacromolecules*, **6**, pp. 612–626.

74. Samir, M., Chazeau, L., Alloin, F., Cavaille, J. Y., Dufresne, A., and Sanchez, J. Y. (2005). POE-based nanocomposite polymer electrolytes reinforced with cellulose whiskers, *Electrochim. Acta*, **50**, pp. 3897–3903.

75. Yun, Y. S., Bak, H., and Jin, H.-J. (2010). Porous carbon nanotube electrodes supported by natural polymeric membranes for PEMFC, *Synth. Met.*, **160**, pp. 561–565.

76. Isikel, L., Gocek, I., and Adanur, S. (2010). Design and characterization of nonwoven fabrics for gas diffusion layer in polymer electrolyte membrane fuel cell, *J. Text. Inst.*, **101**, pp. 1006–1014.

77. Lambert, S., Job, N., D'Souza, L., Pereira, M. F. R., Pirard, R., Heinrichs, B., Figueiredo, J. L., Pirard, J.-P., and Regalbuto, J. R. (2009). Synthesis of very highly dispersed platinum catalysts supported on carbon xerogels by the strong electrostatic adsorption method, *J. Catal.*, **261**, pp. 23–33.

78. Takasu, Y., Ohashi, N., Zhang, X. G., Murakami, Y., Minagawa, H., Sato, S., and Yahikozawa, K. (1996). Size effects of platinum particles on the electroreduction of oxygen, *Electrochim. Acta*, **41**, pp. 2595–2600.

79. Antoine, O., Bultel, Y., Durand, R., and Ozil, P. (1998). Electrocatalysis, diffusion and ohmic drop in PEMFC: Particle size and spatial discrete distribution effects, *Electrochim. Acta*, **43**, pp. 3681–3691.

80. Maillard, F., Pronkin, S., and Savinova, E. R. (2009). Influence of size on the electrocatalytic activities of supported metal nanoparticles in fuel cells related reactions, in *Handbook of Fuel Cells* (Vielstich, W., Gasteiger, H. A., and Yokokawa, H., eds.), John Wiley & Sons, Inc., New York, Vol. 5, pp. 91–111.

81. Maillard, F., Martin, M., Gloaguen, F., and Léger, J. M. (2002). Oxygen electroreduction on carbon-supported platinum catalysts. Particle-size effect on the tolerance to methanol competition, *Electrochim. Acta*, **47**, pp. 3431–3440.

82. Gloaguen, F., and Durand, R. (1997). Simulations of PEFC cathodes: An effectiveness factor approach, *J. Appl. Electrochem.*, **27**, pp. 1029–1035.

83. Wissler, M. (2006). Graphite and carbon powders for electrochemical applications, *J. Power Sources*, **156**, pp. 142–150.

84. Nakamura, K., Shiomi, M., Takahashi, K., and Tsubota, M. (1996). Failure modes of valve-regulated lead/acid batteries, *J. Power Sources*, **59**, pp. 153–157.

85. Hamlen, T. B., and Atwater, R. P. (2002). Metal air Batteries, in *Handbook of Batteries* (Linden, D., and Reddy, T. P., eds.), McGraw-Hill, New York, Chapter 38, pp. 38-1–38-53.

86. Beattie, S. D., Manolescu, D. M., and Blair S. L. (2009). High-capacity Lithium-air cathodes, *J. Electrochem. Soc.*, **156**, pp. A44–A47.

87. Pandolfo, A. G., and Hollenkamp, A. F. (2006). Carbon properties and their role in supercapacitors, *J. Power Sources*, **157**, pp. 11–27.

88. Nagaura, T., and Tozawa, K. (1990). Lithium ion rechargeable battery, *Prog. Batteries Solar Cells*, **9**, pp. 209–217.

89. Flandrois, S., and Simon, B. (1999). Carbon materials for lithium-ion rechargeable batteries, *Carbon*, **37**, pp. 165–180.

90. Prem Kumar, T., Sri Devi Kumari, T., and Manuel Stephan, A. (2009). Carbonaceous anode materials for lithium-ion batteries–the road ahead, *J. Indian Inst. Sci.*, **89**, pp. 393–424.

91. Schlaikjer, C. R. (1983). Lithium oxyhalide batteries, in *Lithium Batteries* (Gabano, J.-P., ed.), Academic Press, London, Chapter 13, pp. 304–363.

92. Marinčić, N. (1975). Materials balance in primary batteries. I. Lithium inorganic cells at low discharge rates, *J. Appl. Electrochem.*, **5**, pp. 313–318.

93. Marinčić, N. (1976). Materials balance in primary batteries. II. Lithium inorganic batteries at high discharge rates, *J. Appl. Electrochem.*, **6**, pp. 51–58.

94. Dey, A. N. (1976). Experimental optimization of Li/$SOCl_2$ primary cells with respect to the electrolyte and the cathode compositions, *J. Electrochem. Soc.*, **123**, pp. 1262–1264.

95. Klinedist, K. A. (1981). Cathode-Limited Li/$SOCl_2$ Cells, *J. Electrochem. Soc.*, **128**, pp. 2507–2512.

96. Szpak, S., and Driscoll, J. R. (1983). Investigation of positive electrode characteristics in high rate Li/$SOCl_2$ cells, *J. Power Sources*, **10**, pp. 343–354.

97. Bagotzky, V. S., Volkovich, Y. M., Kanevsky, L. S., Skundin, A. M., Broussely, M., Chenebault, P., and Caillaud, T. (1995). Changes to the porous structure of carbon cathodes during the discharge of Li-$SOCl_2$ cells, *19th International Power Sources Symposium*, Brighton, pp. 359–372.

98. Carmier, D., Vix-Guterl, C., and Lahaye, J. (2002). Porosity of the cathode during the discharge of $SOCl_2$/Li batteries: Influence of the porous morphology of the carbon used, *J. Power Sources*, **103**, pp. 237–244.

99. Carmier, D., Vix-Guterl, C., and Lahaye, J. (2001). Porosity of the cathode during the discharge of $SOCl_2$/Li batteries: I. Industrial cathodes, *Carbon*, **39**, pp. 2181–2186.

100. Simon, B., Hilaire, M., and Cousseau, J.-F. (2005). Electrochemical generator with an aerogel carbon cathode, European patent EP1610404.

Chapter 6

Environmental and Green Chemistry Applications of Nanoporous Carbons

Juan Matos

Engineering of Materials and Nanotechnology Centre,
Venezuelan Institute for Scientific Research,
20632, Caracas 1020-A, Venezuela

jmatos@ivic.gob.ve, jmatoslale@gmail.com

The influence of surface pH of activated carbon on selectivity of titanium dioxide, (TiO_2) during the photooxidation of 4-chlorophenol (4CP) on UV irradiation was studied as a promising test for verify both the environmental and green chemistry applications of nanoporous carbons. Chlorine substitution by p-hydroxylation yielding hydroquinone in equilibrium with benzoquinone and o-hydroxylation to 4-chlorocathecol were detected as the main intermediates in the photooxidation of 4CP. Changes in the surface pH of carbons from basic to closely neutral induce a remarkable increase in the production of 4-chlorocatechol 22 times higher on TiO_2-AC in comparison with TiO_2 alone. A mechanism of interaction between both of the solids is proposed to explain changes in both the photoactivity and photoselectivity of TiO_2. In conclusion, results showed that TiO_2-AC can be used as an alternative green photocatalyst in organic synthesis.

Green Carbon Materials: Advances and Applications
Edited by Thomas E. Rufford, Denisa Hulicova-Jurcakova, John Zhu

ISBN 978-981-4411-13-4 (Hardcover), 978-981-4411-14-1 (eBook)
www.panstanford.com

6.1 Introduction

An important challenge for industrial organic chemistry is developing chemical processes for a clean environment. This is the field of environmental green chemistry [1, 2]. An interesting alternative for green chemistry processes is heterogeneous photocatalysis, principally, because it directly concerns with several of the 12 principles of green chemistry [3]. Titanium dioxide is a non-toxic and biocompatible material, and it is the most efficient photocatalyst that shows high efficiency for the oxidative photodegradation of hazardous aromatic molecules in polluted water and air. Also, it shows a high selectivity in the functionalization of light n-alkanes, in selective mild oxidation of gas and liquid hydrocarbons, and in selective oxidation of alcohols and hydroxyl-containing molecules [2]. In short, photocatalysis is a promising route for the 21st century organic chemistry [4] where selective photocatalytic conversions will play a major role, offering an alternative green route for the production of organics. However, in aqueous phase, TiO_2 has low selectivity and intermediate products of the photooxidation of aromatic molecules commonly are photomineralized. In previous works, we have showed [5, 6] that TiO_2 photoactivity is strongly influenced by activated carbon surface properties. In this work, the influence of nanoporous carbon's pH on the selectivity of intermediate products formed during the photooxidation of 4CP on UV-irradiated TiO_2 was verified as a possible strategy for highly selective green organic synthesis.

6.2 Experimental

6.2.1 Materials and Characterization

High-purity 4-chlorophenol (4CP) was purchased from Aldrich. The photocatalyst was TiO_2 P25 from Degussa. Activated carbons (AC) were prepared from sawdust of *Tabebuia Pentaphyla* wood by two methods. Pyrolysis under N_2 flow [5] at temperatures from 450 to 1000°C for 1 h (denoted by $AC_{N2\text{-}i}$, where *i* is the activation temperature) and by chemical activation [6] after the impregnation of precursor with different concentrations of H_3PO_4 (from 1% to 65% w/w) followed by activation under N_2 flow at 450°C for 1 h and denoted by $AC_{H3PO4\text{-}i\%}$, where $i\%$ is the concentration of H_3PO_4. The characterization of TiO_2 and AC was performed by BET surface areas (S_{BET}) and surface pH (pH_{PZC}). The full isotherms of N_2 adsorption in the range of 0.03 to 630 Torr were measured in a Micromeritics ASAP-2010 apparatus. The pH_{PZC} values were estimated by the drift method [7]. Fourier transform infra-red (FTIR) experiments were carried out on a spectrophotometer Magna-IR 560 from Nicolette. The powders were mixed with KBr in a 5% (w/w) mixture. The mixed powder was pressed to tablets of 1 cm diameter at 10 ton for 1 min. The transparent tablets were inserted in the apparatus and the spectra were recorded from 4000 to 400 cm^{-1} with a resolution of 5 cm^{-1}. The KBr reference spectrum and CO_2 from ambient conditions have been subtracted to every spectrum in order to verify the functional surface groups presented on activated nanoporous carbons.

6.2.2 Photocatalytic Tests

Experimental conditions of photocatalytic tests have been previously reported [5, 6] but are summarized as follows. Fifty milligram TiO_2 and 10 mg AC were added under stirring in 25 mL at 20°C of 100 ppm (Co = 0.78×10^{-3} mol L^{-1}) of 4CP and maintained in the dark by 80 min to reach adsorption at equilibrium. This adsorption time before irradiation was selected because in previous works we found that after this time 4CP achieve equilibrium of adsorption on $TiO_2\text{-}AC_{N2\text{-}i}$ [5] and $TiO_2\text{-}AC_{H3PO4\text{-}i\%}$ [6]. After this time, the 4CP concentration remains constant in dark conditions [5, 6]. Batch photoreactor was a cylindrical flask (Pyrex, 60 mL) with a bottom optical window of 3 cm diameter and open to air [5]. Irradiation was provided by a high-pressure mercury lamp and IR was filtered by a circulating-water cell equipped with a 340 nm cut-off filter. The photon flux emitted by light source was determined by actinometry using uranyle oxalate as the actinometer and software Logicien Photon version 1.6. The photon flux estimated was 2.9×10^{15} photons cm^{-2} s^{-1}. Disks (0.45 μm) were used to remove particulate matter from aliquots (0.3 mL) before analysis by HPLC with a UV absorbance detector (280 nm). Besides 4CP, hydroquinone (HQ), benzoquinone (BQ), and 4-chlorocatechol (4CT) were detected as the main intermediate products. From kinetic curves of 4CP disappearance as function of irradiation time, apparent first-order rate constants (k_{app}) were obtained as the best kinetic parameter to compare the activity of photocatalysts.

6.3 Results and Discussion

6.3.1 Kinetics of 4CP Disappearance

TiO_2 and AC were added under stirring in 25 mL at 20°C of Co = 0.78 × 10^{-3} mol L^{-1} of 4CP and maintained in the dark by 80 min to reach adsorption at equilibrium. These results have been already reported [5, 6], but in short, only about 10% of 4CP was adsorbed on TiO_2 alone and values adsorbed on different AC were from 10% to 60%. This range of 4CP adsorbed in the dark permits us to follow the kinetics of 4CP photo degradation and therefore to estimate the apparent-constant rates (k_{app}) considering only the photocatalytic process. For example, Figs. 6.1 and 6.2 show the kinetics of 4CP disappearance on TiO_2-AC_{N2} and TiO_2-AC_{H3PO4}, respectively. The results obtained on direct photolysis, neat TiO_2, and some selected nanoporous carbons are also included for comparative purposes.

In the present experimental conditions, direct photolysis was negligible compared with the photocatalytic degradation. This is in concordance with the fact that the lamp flux was cut off only to near-UV and visible light. The disappearance of 4CP under UV-irradiated $AC_{N2\text{-}900}$ and $AC_{N2\text{-}1000}$ in Fig. 6.1 and $AC_{H3PO4\text{-}65}$ in Fig. 6.2 is also negligible, which confirms that AC are photo inactive in 4CP photodegradation, in agreement with previous results [8, 9].

From the kinetic trends of 4CP disappearance (Fig. 6.1), it can be inferred that the photooxidation of 4CP follows an apparent first-order rate, and linear transformation of kinetic results, $\mathrm{Ln}(C_o/C_t) = f(t)$, was employed to estimate apparent first-order rate constants. Table 6.1 shows a summary of the S_{BET}, pH_{PZC} of AC, the 4CP adsorbed in the dark ($4CP_{ads\text{-}dark}$), and the apparent constant rates (k_{app}). Results from Table 6.1 indicate that the surface area and mainly the surface pH of AC clearly influence the TiO_2 photoactivity. It can be pointed out that photoactivity cannot be directly associated with the adsorption of 4CP because from the values presented in Table 6.1, it can be seen that for the specific case of TiO_2-AC_{H3PO4} photocatalysts, the higher the 4CP adsorbed, the lower the k_{app} of 4CP photooxidation. For example, by comparison of k_{app} obtained on TiO_2-AC and that obtained on TiO_2 alone, an interaction factor (I_F) between AC and TiO_2 was estimated, and it can be associated with synergistic or inhibiting effects between both solids when $I_F > 1$ or $I_F < 1$, respectively. The beneficial and detrimental influence of AC on TiO_2 photoactivity has been extensively discussed in previous works and in most of cases has been associated with the presence of a contact interface created between both solids [5, 6, 8–10]. The influence of this interface on photoactivity and selectivity is discussed here.

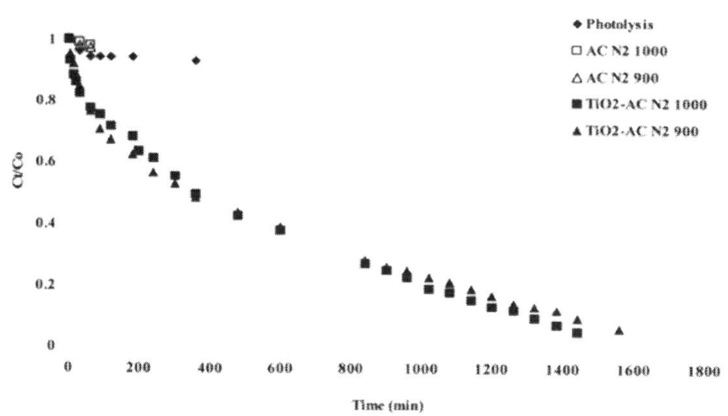

Figure 6.1 Kinetics of 4CP disappearance on TiO_2-AC_{N2} and AC_{N2}.

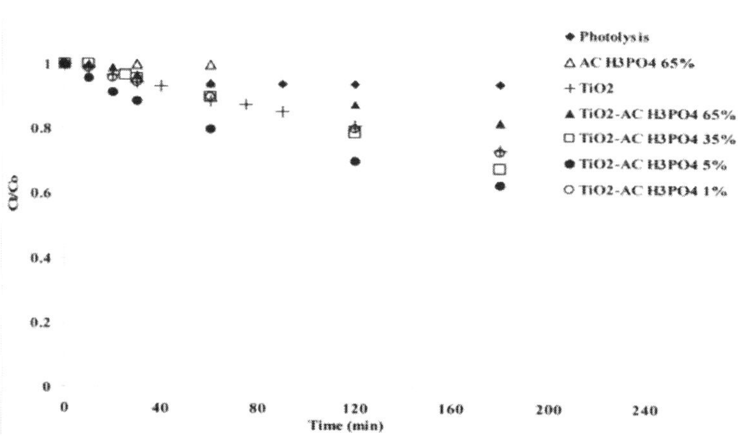

Figure 6.2 Kinetics of 4CP disappearance on TiO_2-AC_{H3PO4} and $AC_{H3PO4\text{-}65\%}$.

Table 6.1 AC properties and kinetic results obtained in 4CP photo-oxidation

Photocatalyst	pH_{PZC}	S_{BET} (m^2/g)	$4CP_{ads\text{-}dark}$ (%)[a]	$k_{app} \times 10^{-3}$ (min^{-1})	I_F[b]
TiO_2	6.5	50	9.8	2.02	1.00
TiO_2-$AC_{N2\text{-}1000}$	8.90	518	23.8	4.31	2.13
TiO_2-$AC_{N2\text{-}900}$	8.50	590	22.5	3.37	1.67
TiO_2-$AC_{N2\text{-}800}$	8.00	519	21.9	1.54	0.76
TiO_2-$AC_{N2\text{-}700}$	7.90	388	6.8	1.34	0.66
TiO_2-$AC_{N2\text{-}600}$	7.10	360	5.8	1.07	0.53
TiO_2-$AC_{N2\text{-}450}$	6.05	31	4.9	0.92	0.46
TiO_2-$AC_{H3PO4\text{-}1\%}$	4.70	188	10.9	2.01	1.00
TiO_2-$AC_{H3PO4\text{-}5\%}$	4.00	414	11.6	3.36	1.66
TiO_2-$AC_{H3PO4\text{-}35\%}$	3.50	1987	51.5	2.09	1.03
TiO_2-$AC_{H3PO4\text{-}65\%}$	3.10	1569	44.1	1.25	0.62

[a]4CP adsorbed in the dark from an initial concentration of 0.78×10^{-3} mol L^{-1} (100 ppm).
[b]Synergy or inhibition described by: $I_F = [k_{app}(TiO_2\text{-}AC)/k_{app}(TiO_2) + k_{app}(AC\text{-}i)]$.

6.3.2 Kinetics of Appearance and Disappearance of Intermediate Products

The influence of surface pH of AC upon the kinetics of 4CP photodegradation (Table 6.1) was confirmed by the fact that though intermediate products generated during the degradation of 4CP are not quite different, according to the type of carbon materials employed, there is a clear change in the intermediate product distribution. Figure 6.3 shows the kinetic trends corresponding to the sum of hydroquinone (HQ) and benzoquinone (BQ) and Fig. 6.4 shows the corresponding one to 4CT detected on irradiated TiO_2 in presence of the acid carbons (AC_{H3PO4}).

It can be seen in Figs. 6.3 and 6.4 that for TiO_2 alone the main intermediate products are HQ + BQ (Fig. 6.2). Figure 6.3 also shows that the more acid is the AC employed, the higher is the production of intermediate products. Figure 6.4 clearly shows that the 4CT formation is clearly higher in the presence of the most acid TiO_2-AC_{H3PO4} photocatalysts (65%, 35%, 5% w/w) and 4CT was practically negligible for the TiO_2-$AC_{H3PO4\text{-}1\%}$ and for TiO_2 alone. This indicates that different surface sites governed by pH_{PZC} lead to different types of 4CP adsorption mechanisms with a clear influence on the intermediate product distributions. Table 6.2 shows a summary of these result products detected in the photooxidation of 4CP on TiO_2-AC.

Table 6.2 Intermediate products detected in the photooxidation of 4CP

Photocatalyst	HQ + BQ (μmol)	4CT (μmol)	*R*(o/p)	*F*(o/o)
TiO_2	0.476	0.037	0.08	1.00
TiO_2-$AC_{N2\text{-}1000}$	0.487	0.059	0.12	1.60
TiO_2-$AC_{N2\text{-}900}$	0.300	0.068	0.23	1.84
TiO_2-$AC_{N2\text{-}800}$	0.210	0.680	3.24	18.4
TiO_2-$AC_{N2\text{-}700}$	0.139	0.817	5.88	22.1
TiO_2-$AC_{N2\text{-}600}$	0.135	0.805	5.96	21.8
TiO_2-$AC_{N2\text{-}450}$	0.106	0.790	7.45	21.4
TiO_2-$AC_{H3PO4\text{-}1\%}$	0.363	0.108	0.30	2.92
TiO_2-$AC_{H3PO4\text{-}5\%}$	0.381	0.493	1.29	13.3
TiO_2-$AC_{H3PO4\text{-}35\%}$	0.442	0.540	1.22	14.6
TiO_2-$AC_{H3PO4\text{-}65\%}$	0.690	0.642	0.93	17.4

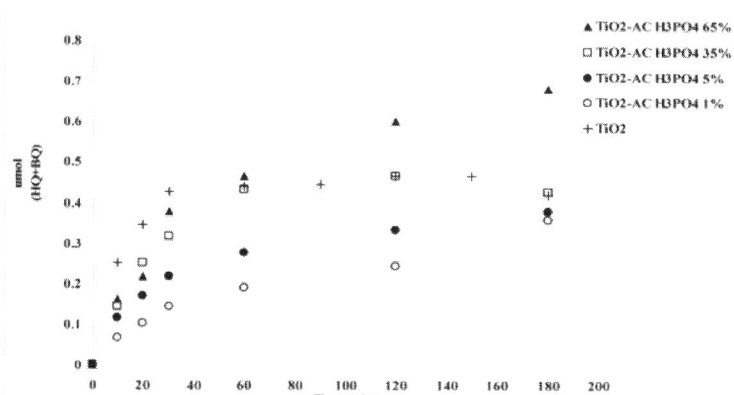

Figure 6.3 Kinetics of appearance and disappearance of HQ and BQ on TiO_2-AC_{H3PO4}.

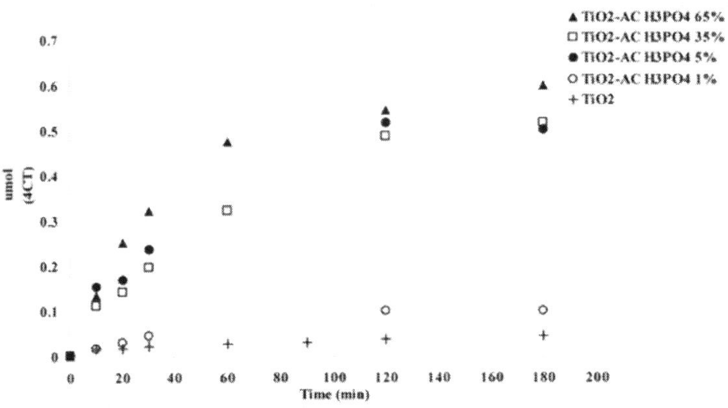

Figure 6.4 Kinetics of appearance and disappearance of 4CT on TiO_2-AC_{H3PO4}.

Intermediate products detected are in good agreement with chlorine substitution by p-hydroxylation yielding HQ and BQ that are commonly in equilibrium with o-hydroxylation yielding 4CT. Table 6.2 shows a summary of changes detected in ortho/para products in the 4CP photooxidation. *R*(o/p) compares the maximum quantity detected of 4CT against maxima of HQ and BQ. Values in Table 6.2 indicate a clear increase of *R*(o/p) in comparison with that obtained on TiO_2 alone (0.08). *R*(o/p) yields a maxima value of about 7.453 for TiO_2-$AC_{N2\text{-}450}$ (pH_{PZC} 6.05). A similar trend was detected for the relation *F*(o/o), which compares the maxima quantity of 4CT detected on TiO_2-AC against maxima 4CT on TiO_2 (0.037 μmol). It can be seen in Table 6.2 that for the case of $AC_{N2\text{-}700}$ against $AC_{N2\text{-}1000}$, a light change in the surface pH (from 8.9 to 7.9) leads to a remarkable increase in *F*(o/o) from 1.60 up to a maxima of about 22.1 for TiO_2-$AC_{N2\text{-}700}$. A similar behavior is detected for the other type of AC, which is clearly more acid than those prepared by pyrolysis. In addition, bearing in mind that 4CP photooxidation occurs on the TiO_2 surface, the present results clearly show that light changes in pH_{PZC} of AC induce important changes in the surface of the semiconductor and this clearly affects the reaction mechanism. Figures 6.5 and 6.6 show the influence of pH_{PZC} of AC_{N2} and AC_{H3PO4}, respectively, on the *R* and *F* parameters. It can be seen in both cases that the lower the pH_{PZC} of AC, the higher the values of *R* and *F* indicated in Table 6.2.

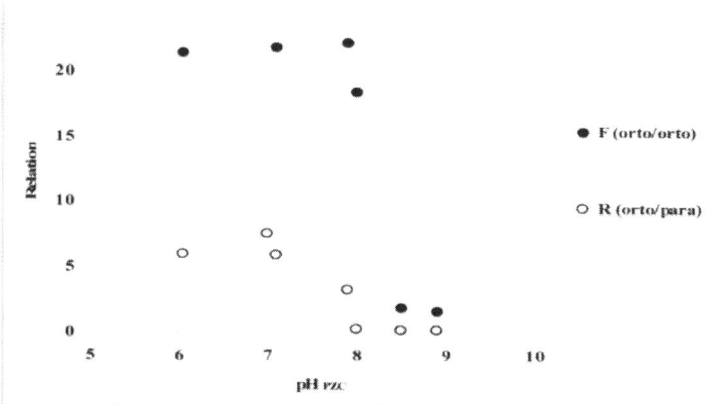

Figure 6.5 Influence of pH_{PZC} of AC_{N2} on selectivity.

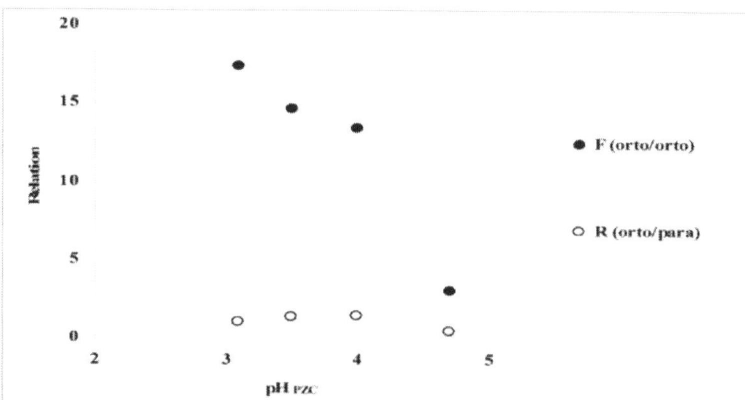

Figure 6.6 Influence of pH_{PZC} of AC_{H3PO4} on selectivity.

In the photooxidation of 4CP, there are two possible reaction mechanisms. First, the oxidation of 4CP can occur by the nucleophilic substitution of chlorine by the attack of the aromatic ring with °OH radicals that are promoted at neutral or basic pH. This route yields preferential HQ and its successive oxidation yields BQ. On the other hand, the oxidation of 4CP would occur by the oxidation of aromatic ring by electrophilic addition of anion radical $O_2^{\circ-}$ yielding higher selectivity to 4CT. This mechanism is favored at acid pH. These results are in concordance with previous studies regarding the selective photooxidation of phenol [8] and 2,4-dichlorophenoxiacetic acid [9].

These remarkable differences in products selectivity as a function of AC's pH_{PZC} can be associated with the detection of an interfacial interaction between both solids. For example, infrared spectroscopy showed that some functional organic groups are presented on AC's surface, principally cyclic ethers for basic AC [5, 10] and carboxylic acids in acid AC [6]. In those works, we have found that these organic groups are responsible of the solid–solid interaction by means of the coordination of the oxygenated functional groups on AC surface [6] and the metallic centre (Ti^{4+}) in TiO_2. This suggested interaction agrees with the oxofunctionalization reported with other kind of photocatalyst, as silica-supported polyoxometalates in organic selective synthesis [11] and with the evidence of electron transfer from AC to TiO_2 reported by Godínez [12]. An interaction mechanism between TiO_2 and acid AC is proposed in Fig. 6.7. The case shown in Fig. 6.7a corresponds to the transfer of electrons from the conduction band of TiO_2 to AC with a more basic surface pH and therefore with semiconductor behavior.

On the other hand, the mechanism shown in Fig. 6.7b involves the transfer of electrons from AC to TiO_2 by the coordination of carboxylic acid groups from AC's surface to Ti^{4+} centers in TiO_2. In the case of the transfer of electrons from Titania to AC (Fig. 6.7a), this kind of interaction permits us to increase °OH radicals in agreement with the strong oxidation of 4CP by means of a mechanism to mainly produce para-products (Table 6.2) such as HQ and BQ. This p-hydroxylation mechanism is induced by basic surface pH of AC, and it involves the abstraction of a hydride and chloride from 4CP to produce benzene (Fig. 6.8), which is readily oxidized to produce HQ, commonly in equilibrium with BQ. On the other hand, the interaction between TiO_2 and AC shown in Fig. 6.7b promotes higher concentrations of O_2^{o-} and HO_2^{o} (obtained from O_2^{o-} + H^+) radicals whose induce remarkable changes in selectivity of products to mainly produce ortho-product as 4CT. This o-hydroxylation mechanism is induced by acid surface pH of AC, and it involves the abstraction of an electron from 4CP to produce an aromatic cation radical (Fig. 6.9) that is stabilized by resonance to produce the structure 6C, which is readily oxidized to produce selectively 4CT and H_2O_2, a common sub-product obtained in the present experimental conditions.

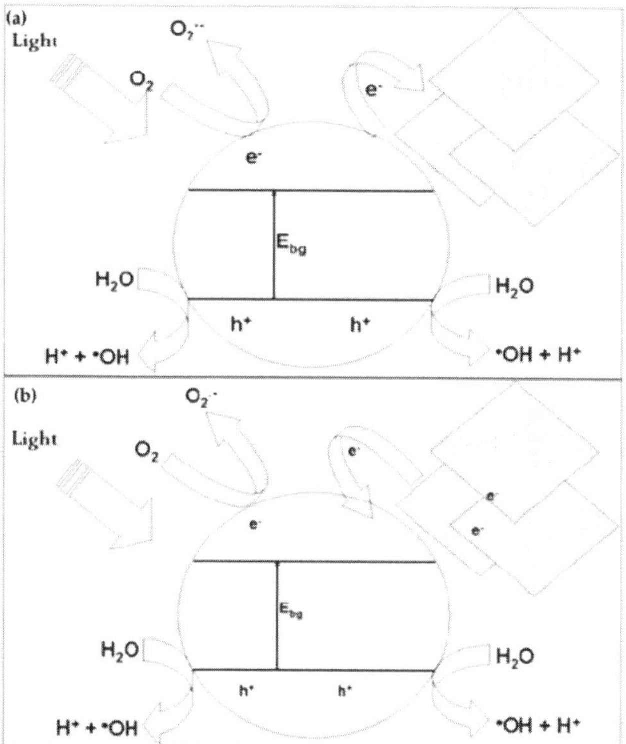

Figure 6.7 Scheme of interaction between TiO_2 and AC. (a) Basic AC, (b) acid AC.

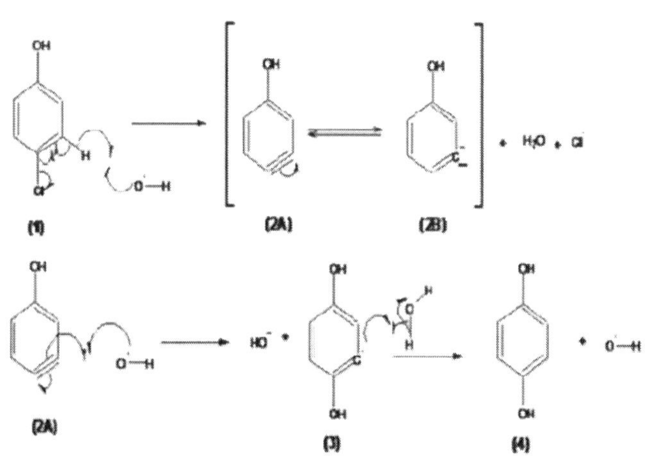

Figure 6.8 Mechanism of p-hydroxylation induced by basic surface pH of AC.

Accordingly, the interpretation of pH effects on the selective photooxidation of 4CP is a very difficult task because the pH of the solution influences the adsorption and dissociation of substrate, catalyst surface charge, oxidation potential of the valence band, and other physicochemical properties of the system [13, 14].

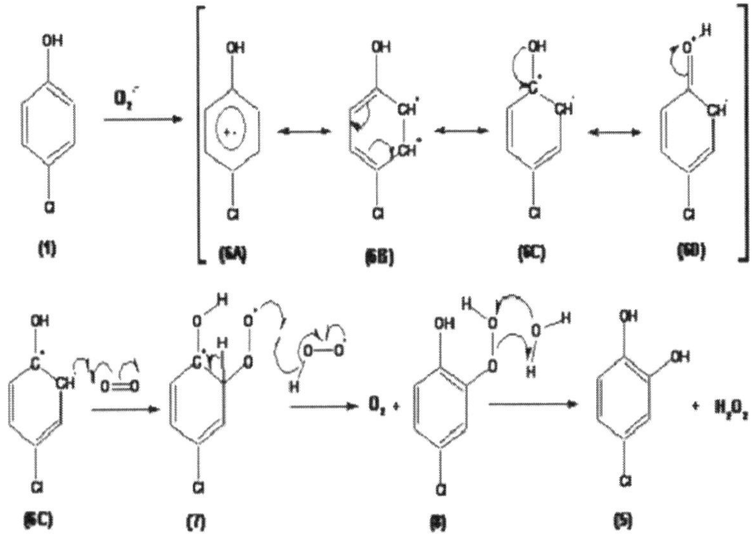

Figure 6.9 Mechanism of o-hydroxylation induced by acid surface pH of AC.

6.4 Conclusions

Changes in pH_{PZC} of AC allow the control of the selectivity of products during 4CP photooxidation on TiO_2-AC. The use of several advanced techniques such as XPS, XANES, EXAFS, and photoelectronic conductivity is required for a better understanding of TiO_2-AC binary materials in order to properly establish the most important characteristics of nanoporous AC on the selective photooxidation of 4CP and other aromatic molecules. In spite of this, results suggest the possibility of the use of TiO_2-AC as an alternative environmental green photocatalyst with high selectivity in organic synthesis.

References

1. Anpo, M. (2000). Utilization of TiO_2 photocatalysts in green chemistry. *Pure Appl. Chem.*, **72**, 1265–1272.
2. Herrmann, J. M., Duchamp, C., Karkmaz, M., Hoai, B. T., Lachheb, H., Puzenat, E., and Guillar, C. (2007). Environmental green chemistry as defined by photocatalysis. *J. Hazard. Mater.*, **146**, 624–629.
3. Anastas, P. T., and Warner, J. C. (1998). *Green Chemistry: Theory and Practice*. Oxford University Press, New York. p. 30.
4. Palmisano, G., Augugliaro, V., Pagliaro, M., and Palmisano, L. (2007). Photocatalysis: a promising route for 21st century organic chemistry. *Chem. Commun.*, **33**, 3425–3437.
5. Cordero, T., Duchamp, C., Chovelon, J. M., Ferronato, C., and Matos, J. (2007). Surface nano-aggregation and photocatalytic activity of TiO_2 on H-type activated carbons. *Appl. Catal. B Environ.*, **73**, 227–235.
6. Cordero, T., Duchamp, C., Chovelon, J. M., Ferronato, C., and Matos, J. (2007). Influence of L-type activated carbons on photocatalytic activity of TiO_2 in 4-Chlorophenol photodegradation. *J. Photochem. Photobiol. A Chem.*, **191**, 122–131.
7. Lopez, M. V., Stoeckli, F., Moreno, C., and Carrasco, F. (1999). On the characterization of acidic and basic surface sites on carbons by various techniques. *Carbon*, **37**, 1215–1221.
8. Matos, J., Laine, J., and Herrmann, J. M. (1998). Synergy effect in the photocatalytic degradation of phenol on a suspended mixture of Titania and activated carbon. *Appl. Catal. B Environ.*, **18**, 281–291.
9. Matos, J., Laine, J., and Herrmann, J. M. (2001). Effect of the type of activated carbons on the photocatalytic degradation of aqueous organic pollutants by UV-irradiated Titania. *J. Catal.*, **200**, 10–20.
10. Matos, J., Laine, J., Herrmann, J. M., Uzcategui, D., and Brito, J. L. (2007). Influence of activated carbon upon Titania on aqueous photocatalytic consecutive runs of phenol photomineralization. *Appl. Catal. B Environ.*, **70**, 461–469.
11. Maldotti, A., Molinari, A., and Amadelli, R. (2002). Photocatalysis with organized systems for the oxofunctionalization of hydrocarbons by O_2. *Chem. Rev.*, **102**, 3811–3836.
12. Peralta-Hernández, J. M., Manríquez, J., Meas-Vong, Y., Rodríguez, F. J., Chapman, T. W., Maldonado, M. I., and Godínez, L. A. (2007).

Photocatalytic properties of nano-structured TiO_2-carbon films obtained by means of electrophoretic deposition. *J. Hazard. Mater.*, **147**, 588–593.

13. Matos, J., García, A., Cordero, T., Chovelon, J. M., and Ferronato, C. (2009). Eco-friendly TiO_2-AC photocatalyst for the selective photooxidation of 4-chlorophenol. *Catal. Lett.*, **130**, 568–574.
14. Matos, J., García, A., and Poon, P. S. (2010). Environmental green chemistry applications of nanoporous carbons. *J. Mater. Sci.*, **45**, 4934–4944.

Chapter 7

Removal of Toxic Gases by Activated Carbons

Teresa J. Bandosz

Department of Chemistry, The City College of New York, 160 Convent Ave., New York, NY 10031, USA

tbandosz@ccny.cuny.edu

7.1 Introduction

Various technological processes result in the emission of toxic gases into the atmosphere. These gases are classified as toxic industrial compounds (TICs), and efforts are made to decrease/eliminate their presence in air. Besides causing detrimental environmental changes such as acid rains, photochemical smog, and global warming [1, 2], these gases can also cause a direct threat to human health when they are released as a result of terrorists actions.

The Clean Air Act of the US government introduced in the early 1990s [2, 3] resulted in a significant decrease in the amount

Green Carbon Materials: Advances and Applications
Edited by Thomas E. Rufford, Denisa Hulicova-Jurcakova, John Zhu

ISBN 978-981-4411-13-4 (Hardcover), 978-981-4411-14-1 (eBook)
www.panstanford.com

of gases emitted into the atmosphere. Also, new technologies were engaged for the desulfurization of fuel, cleaning of stock gases, and improvement in the efficiency of combustion.

One of the most promising technologies that have led to a significant decrease in the air concentration of toxic gases is adsorption on activated carbons. These materials are known for their large surface area, high porosity, and catalytic properties. They are obtained using relatively inexpensive technology. Moreover, their surface can be modified using various methods.

This chapter summarizes the recent efforts for the removal of TICs such as ammonia, hydrogen sulfide, sulfur dioxide, nitric oxides, and hydrogen cyanide with activated carbons as the adsorbents of choice. The focus is on surface science in involved adsorption/reactive adsorption process. Therefore, the combined role of heteroatoms, inorganic matter, and porosity in the enhancement of surface interactions/reactions is emphasized.

7.2 Removal of Ammonia

Ammonia is one of the most widely used chemicals [4] and thus an important source of nitrogen compounds in the atmosphere [1]. Generally, ammonia adsorption can take place following two different processes: physisorption and chemisorption [5, 6]. To be effective, it must involve specific interactions and chemical transformations such as hydrogen bonding, acid–base reactions (ammonia, basic molecule), complexation (ammonia, common ligand), and others [7].

Even though activated carbons are versatile adsorbents [8–10], their use in ammonia retention is limited. Because the size of ammonia molecule is rather small (3 Å, [11]) and it is a very basic molecule, for its efficient adsorption an adsorbent with very small pores and an acidic surface is required. Average activated carbons are not good candidates for this process since they have pores with sizes between 10 and 20 Å and their surface is rather basic [12]. Therefore, various methods have been applied to modify activated carbons toward an increased efficiency in ammonia removal. One of the most common methods is surface oxidation [6, 13–16]. Another approach involves impregnation/deposition on the carbon surface such species as metal oxides [7], acids and bases [17], or heteropolyacids [18]. Owing to the solubility of ammonia, the effects of water on its retention have also been examined [19, 20] because on the large scale the presence of humidity is difficult and expensive to avoid.

Petit and coworkers studied extensively ammonia adsorption under dynamic conditions, at room temperature, on activated carbons of different origins (coal-based, wood-based and coconut shell-based carbons). The carbons were used as received or were impregnated with various inorganic compounds, including metal chlorides [21], metal oxides [22], and polycations [23]. The tests were run in both dry and moist conditions to evaluate the effects of moisture. The breakthrough tests results showed significant differences in ammonia adsorption capacity depending on the carbon type, the impregnants, and the experimental conditions. Surface chemistry was indicated as the main factor governing ammonia adsorption on the impregnated carbons. A proper combination of the surface pH, the strength, type, and the amount of functional groups present on the adsorbents' surface determines the ammonia uptake. Interestingly, water can either enhance or inhibit ammonia retention [19]. The former happens owing to ammonia dissolution in water, formation of ammonium ions, and their reactions with carboxylic groups from the carbons' surface [6, 24–28]. When the strength of ammonia adsorption is considered, the presence of water has a negative effect since ammonia dissolved in water is first removed from the system when the adsorbent bed is purged with air. Ammonia, besides adsorption by van der Waals forces and dissolution in water, is also retained on the surface *via* reactive mechanism, which involves acid–base reactions (Brønsted and Lewis) or complexation [6, 19–23, 27–29]. The results of Petit and coworkers' study indicated that 6–47% ammonia adsorbed is strongly retained on the surface even when the bed is purged with air (Fig. 7.1) [29].

Stoeckli and coworkers suggested that the adsorption of ammonia on activated carbons is governed by two mechanisms that follow the Type I DR isotherms [18]. The first mechanism represents specific interactions and the amount adsorbed was found to be governed by the amount of oxygen present on the carbons surface. As a second mechanism, the simple micropore filling was proposed.

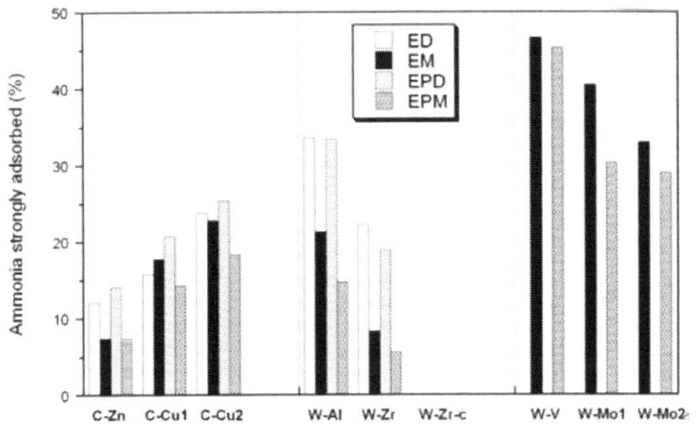

Figure 7.1 Evaluation of the strength of ammonia retention on the surface of the modified samples in dry (ED), moist (EM) conditions. P refers to 2 h prehumidification procedure. Reprinted with permission from Ref. [29]. Copyright Elsevier, 2009.

Interesting results on ammonia adsorption were obtained on sulfur-containing nanoporous carbons [28]. It was found that not only oxygen-containing groups but also sulfur-containing groups enhance ammonia adsorption. Sulfonic groups incorporated to carbon matrix in the presence of superoxide anions are converted into sulfates that react with ammonia to form ammonium sulfates. These salts are strongly retained in the micropores of the adsorbents.

Much better adsorbents of ammonia at ambient conditions than activated carbons are graphite oxides [19, 20, 27]. On them, the capacity over 10 wt% was reached. Moreover, the adsorption was strong owing to the reactivity of graphite oxide functional groups with ammonia and strong hydrogen bonding with the interlayer space. Water enhances the amount of NH_3 adsorbed *via* its dissolution. It also promotes the dissociation of surface functional groups. This enhances the formation of ammonium ions. The negative effect of water is in its screening actions on the accessibility of epoxy and -OH groups for reaction with ammonia [19]. The adsorption of NH_3 on a partially dried graphite oxide was indicated as being enhanced not only due to reactive adsorption but also due to the formation of new adsorption centers as a result of the incorporation of ammonia to graphene layers [30].

The following reactions, which contribute significantly to the reactive adsorption of ammonia on graphite oxide/activated carbons containing oxygen and/or sulfur functional groups were proposed [19, 20, 27] in which either ammonium salts or amides are formed:

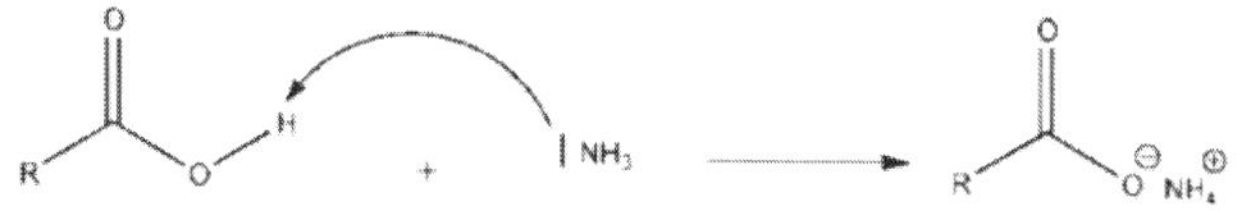

(7.1)

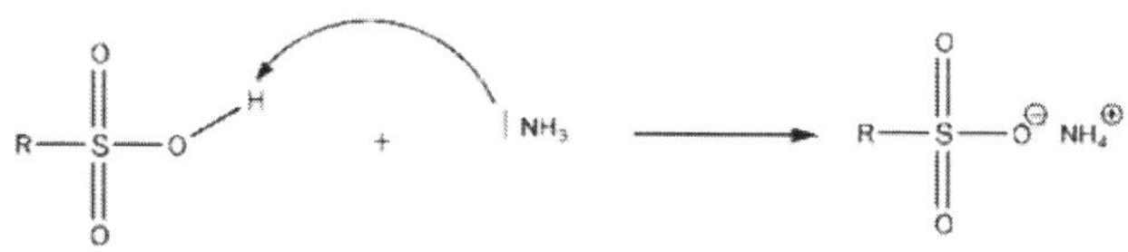

(7.2)

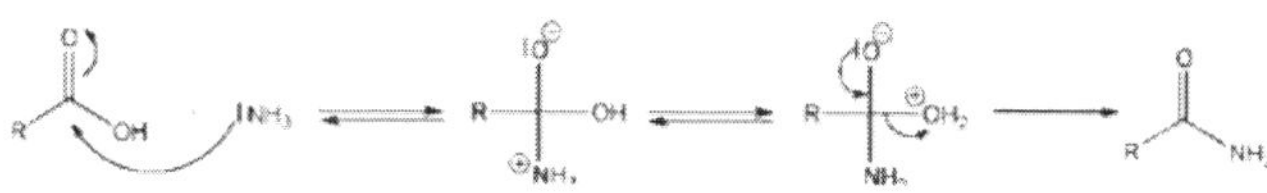

(7.3)

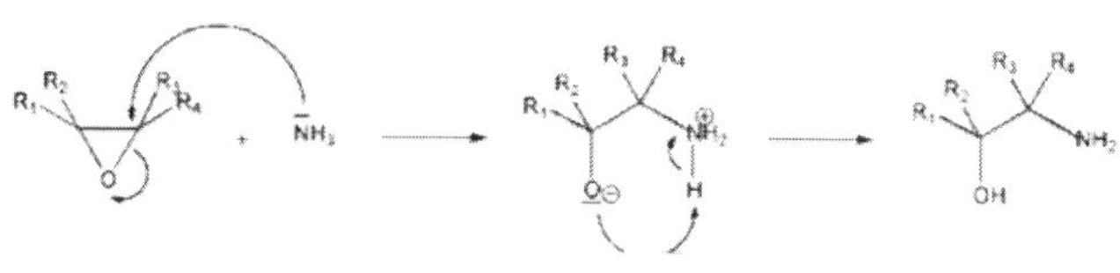

(7.4)

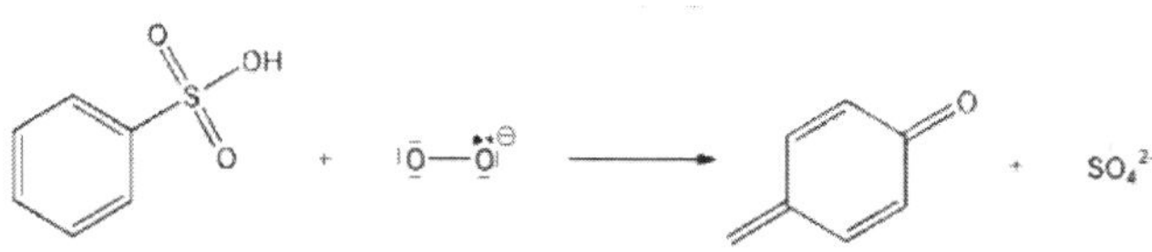

(7.5)

$$2NH_4^+ + SO_4^{2-} \longrightarrow (NH_4)_2SO_4 \quad (7.6)$$

The effect of the impregnation of the carbons with various transition metal salts on the adsorptive behavior has been studied in detail by Dahn and coworkers [31, 32]. They found that the way the salts deposits on the surface significantly affects the performance. This is related to the hygroscopic behavior of salts. Some salts, such as $ZnCl_2$ in monolayer form, were found to be less hygroscopic than that in the bulk phase or aggregates. Since water generally enhances the adsorption of ammonia, the hydrophilic character of salts is a plus. On the other hand, the formation of aggregates blocks the porosity and limits the adsorption capacity.

To increase the ammonia adsorption capacity, hybrids of activated carbon with minerals such as sepiolite were also proposed [33]. Even though the sepiolite components increase the specific interactions of ammonia with the surface, its main role is in binding the carbon particles.

7.3 Removal of Hydrogen Sulfide

Hydrogen sulfide is one of the most potent TICs with a strong environmental impact. Besides its high toxicity and strong odor, it causes acid rain. Since it is an acidic gas, activated carbons impregnated with NaOH or KOH are often used for its removal from air in water treatment plants [34–38]. Water in the air stream is beneficial for the surface reactions [39–41]. To evaluate the H_2S removal capacity of caustic carbons, accelerated ASTM D6646-01 test is used [42]. It is recommended that a good carbon used to clean air in water treatment plant should be able to remove 0.140 g of H_2S per cm^3 of carbon bed. Caustic-impregnated carbons also exhibit catalytic oxidation properties [41] engaging at least 3 moles of H_2S per 1 mole of NaOH. This is owing to a beneficial shift in H_2S dissociation and an increase in the formation of HS^- ions, which are oxidized to elemental sulfur. Therefore, on the surface of such carbons besides sulfur, sulfides, sulfates and carbonates are formed. The disadvantage of caustic carbons' application is their low ignition temperature [37, 38] and formation of elemental sulfur [37, 38, 41], which means that these adsorbents cannot be regenerated *in situ* by washing with water [43]. The surface reactions paths on caustic carbons are expressed as follows [41]:

$$NaOH + H_2S \rightarrow NaHS + H_2O \tag{7.7}$$

$$2NaOH + H_2S \rightarrow Na_2S + H_2O \tag{7.8}$$

$$NaHS + 0.5O_2 \rightarrow S + NaOH \tag{7.9}$$

$$Na_2S + 0.5O_2 + H_2O \rightarrow S + 2NaOH \tag{7.10}$$

$$HS^- + H_2O \rightarrow H_2S + OH^- \tag{7.11}$$

$$S^{2-} + H_2O \rightarrow HS^- + OH^- \tag{7.12}$$

$$2NaOH + H_2SO_4 \rightarrow Na_2SO_4 + 2H_2O \tag{7.13}$$

Another impregnant that increases the capacity of activated carbon to remove hydrogen sulfide is potassium carbonate [44, 45]. It facilitates the dissolution of H_2S that deposits in micropores as KHS. The latter decomposes to H_2S and in this form is oxidized to elemental sulfur in small pores. As in the case of caustic-impregnated carbons, oxidation releases a significant amount of heat.

The low ignition temperature of impregnated carbons directed the research focus to virgin (unimpregnated) activated carbons [46–78] or carbon with specific surface modifications leading to an increase in their number of basic groups, such as nitrogen functional groups [79–86]. Even though these carbons exhibit lower capacities than that of the caustic counterparts, they are less expensive and the risk of self-ignition is much lower. The majority of research reported on these materials published in the 1980s and 1990s addressed the specific type of carbons without a generalization of the adsorption mechanism [47–78].

According to the mechanism proposed by Hedden and co-workers [60], the dissociation of hydrogen sulfide occurs in the film of adsorbed water on the carbon surface and HS^- ions are oxidized by oxygen radicals to elemental sulfur. The importance of water [55, 61, 63, 65, 67, 69, 77, 78], role of oxygen [47–56], autocatalysis by sulfur [56, 57], influence of pore sizes [48, 58, 62, 64, 67], role of carbon surface chemistry [70–73], the effects of ash [78, 87–89], and speciation of surface oxidation products [70–75] have been extensively analyzed.

The capacity in dry conditions is usually small, and it represents mainly physical adsorption in the small pores of carbons. The adsorption of water on a predominantly hydrophobic carbon surface [90] can be enhanced by functional groups containing oxygen or nitrogen (heteroatoms with the ability of hydrogen bonding) existing at the edges of graphene-like sheets [91, 92]. The prehumidification of the carbon surface before exposure to H_2S was reported to improve the capacity as much as 80 times [78]. On the other hand, too much water can cause its condensation in small pores and limit the contact of HS^- with the carbon surface. It was suggested that the carbon should contain only up to 5% of water [70, 78] to reach the maximum capacity. Since the affinity of carbon to retain water is related to its acidity [70–73, 91, 92], too acidic surface can have an adverse effect on the dissociation of hydrogen sulfide.

Another important aspect of H_2S adsorption on activated carbons is the role of oxygen in this process [47–56, 58]. The results suggested that optimum temperature for high H_2S conversion with low SO_2 production is 448 K with O/H_2S ratio 10.05 times the stoichiometric ratio [52]. The adsorption of oxygen and/or hydrogen sulfide from the bulk phase on the activated carbon surface was identified as a rate-limiting step for the oxidation reactions. At elevated temperatures, oxygen was chemisorbed (73.8 kJ/mol), whereas the physical adsorption of H_2S was suggested [50]. At temperatures less than 400 K, in the absence of air the heat of H_2S adsorption ranged from 40 and 50 kJ/mol and it was found to depend only on the pore sizes [76].

The autocatalysis of H_2S oxidation by deposited sulfur is another interesting aspect of the hydrogen sulfur removal process [56, 57]. Whereas small amounts of deposited sulfur as sulfur radicals at the beginning of the removal process were found to increase the catalytic activity of the carbon, excess sulfur was related to the porosity blocking and detrimental effect on surface reactivity. It was estimated that the catalytic activity per square meter of total surface area is approximately proportional to the amount of adsorbed sulfur. On the other hand, no evidence of sulfur autocatalysis was found by Ghosh and Tollefson [50, 51].

For the effective H_2S adsorption on activated carbons, the presence of micropores is of paramount importance and the filling of micropores by elemental sulfur or sulfides ceases the adsorption process [70, 74, 75, 77]. The strongest sulfur adsorption takes place in pores with sizes between 0.5 and 1 nm [48]. They are similar in sizes to sulfur chain/ring and thus the adsorption potential is overlapped there. On the other hand, in smaller pores, isolated adsorbed sulfur radicals are further oxidized to SO_2 and then to SO_2 resulting in the deposition of sulfuric acid in moist conditions [70]. Support for the importance of porosity is the fact that normalization of H_2S capacity to the pore volume resulted in comparable capacity values (Fig. 7.2) for the large sample of carbons studied [70, 77]. Besides working as nanoreactors, the pores play a significant role in storing H_2S oxidation products [93].

Since hydrogen sulfide undergoes oxidation reactions on the surface of carbons, their chemistry should not be neglected. Linking the acidic properties of hydrogen sulfide to the performance of carbons showed the dependence of the capacity on the chemical nature of the carbonaceous adsorbent [67, 70, 77]. Establishing the threshold in the surface pH beneficial for H_2S oxidation over 4.5 [77] clearly showed that the local pH in the pore system has a significant effect on oxidation to various sulfur species (Fig. 7.2). Thus, the H_2S oxidation path is related to its degree of dissociation. At moderately low average pH, the dissociation of H_2S is limited and a small amount of hydrogen sulfide ions is formed. Since those ions are adsorbed in small pores without formation of polymer, they are quickly oxidized to sulfur oxides and in the presence of water oxysulfur acids are formed. High pH in the basic range shifts the dissociation of H_2S to the right. This process favors formation of polysulfide and bulky sulfur polymers resistant to oxidation. At very acidic pH, the physical adsorption of H_2S is a predominant retention mechanism. Therefore, the reactive adsorption of H_2S can be described as following four steps [70, 77]:

(1) the adsorption of hydrogen sulfide molecules on the carbon surface,
(2) their dissolution in a water film
(3) the dissociation of H_2S in water film
(4) the oxidation of HS^- ions by adsorbed oxygen or formation of polysulfides and sulfur polymers

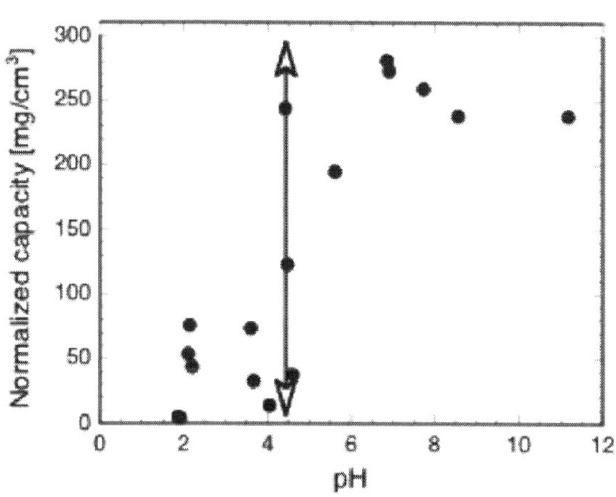

Figure 7.2 Dependence of the normalized H_2S breakthrough capacity on the surface pH. Reprinted with permission from Ref. [77]. Copyright Elsevier, 2001.

The fourth mechanism is dependent on the pH value and on the sizes of pores. This mechanism is considered a pH-dependent mechanism of H_2S oxidation. To summarize, the results of the complex process of oxidation and its yield depend on the specific environment inside the porous structure of carbon. This environment consists of the combination of pore sizes, degree of activated carbon hydrophilicity (related to surface functional groups), and local pH.

This mechanism also governs the H_2S removal on nitrogen-containing carbons [79–86]. An example is Centaur® [82], which features a surface with nitrogen in pyridinic or quaternary configuration [85]. These basic species in the presence of moisture enhance hydrogen sulfide dissociation, adsorption, oxidation, formation of radicals and then their oxidation to sulfuric acid. Even though Centaur® is not considered the best capacity carbon (around 0.060 g/cm^3) [86], its surface conversion of hydrogen sulfide to sulfuric acid is almost complete and therefore its regeneration using simple water washing is feasible [79, 80].

Besides functional groups, the ash components can affect the surface pH and thus surface reactions. The presence of iron oxides or metals ions from group 6–8 has an effect on hydrogen sulfide adsorption [48, 78, 87, 88]. They promote the formation of SO_2 when removal process occurs either at elevated [48] or at ambient temperatures [61, 78]. This effect was also clearly seen on sewage sludge derived adsorbents when they were tested as hydrogen sulfide removal media [94]. Besides iron, copper and zinc oxides are active in reactions with H_2S; however, the activity diminished when the temperature of process was lowered [89].

Another interesting modification of carbonaceous materials toward an improvement in H_2S removal efficiency is addition of magnesium or calcium oxides to the microporous carbon phase. Midas® and Darco H_2S® were manufactured using this route. On the highly dispersed basic inorganic phase in the presence of water, the adsorption of H_2S takes place, and HS^- ions are oxidized to elemental sulfur and move to small carbon pores, which are located in the close vicinity of catalytic centers [95–97]. The catalytic action takes place until all pores are filed with sulfur. Such carbons reach 60 wt% capacity for hydrogen sulfide removal. The drawback is that CO_2 present in air can poison the catalytic phase [95, 96]. The schematic interpretation of the adsorption process on such carbons is presented in Fig. 7.3.

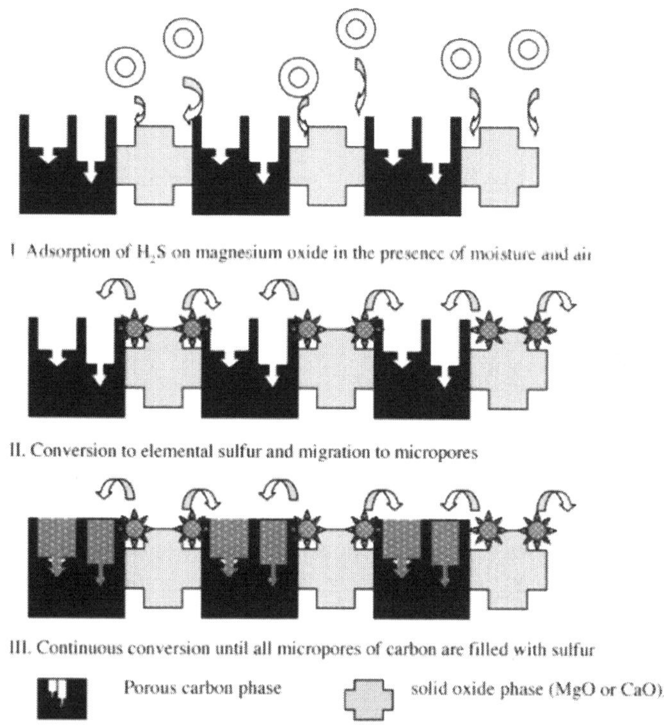

Figure 7.3 Schematic representation of the processes occurring on the surface of adsorbents consisting of a mixture of activated carbon phase and catalytic inorganic phase. Reprinted with permission from [Ref. 95]. Copyright American Chemical Society, 2005.

The effects CO_2 and CH_4 present in challenge gas on H_2S adsorption were also analyzed [95, 96]. The results indicated that when the process is based on caustic materials or alkaline earth oxide, CO_2 has very negative effects on the capacity owing to its reactivity with the bases. Methane may also decrease the capacity when the small pores being able to physically adsorb this molecule are present. Generally, the H_2S removal capacities from air reported on activated carbons and measured using ASTM D6646-01 method [42, 98] vary from 0.003 to 0.200 g/cm^3 [9] depending on their surface modifications and extent of the catalytic properties. This test results limits the data to the ambient conditions. Removal of H_2S from syngas at high temperature is based mainly on catalytic processes and it is beyond the scope of this review.

7.4 Removal of Sulfur Dioxide

Another sulfur-containing species that has detrimental effects on the environment is sulfur dioxide. Efficient media for sulfur dioxide removal are activated carbons [99–117] and activated carbon fibers [105, 111, 115]. As for hydrogen sulfide, for SO_2 adsorption, parameters such as porosity [101–110, 118, 119], surface chemistry [101, 104, 106, 108, 110, 111, 113, 115], and constituents of ash [120–122] are important surface features. Since sulfur in SO_2 is at a higher oxidation state than in H_2S, the chemistry involved in its immobilization is less complex than that taking place during the oxidation of hydrogen sulfide. When the process is carried out in the presence of moisture and oxygen, sulfur dioxide is oxidized to SO_3 and then to sulfuric acid, which is strongly retained in the pore system of activated carbons. Higher extent of oxidation was indicated as resulting in more SO_2 adsorbed [109].

It was demonstrated that oxidation to sulfur trioxide occurs mainly in the 0.7 nm pores [109, 110, 118, 119]. Owing to an increase in the pore sizes, less SO_2 is adsorbed. Interestingly, in the absence of oxygen, the pore volume plays a more important role than pore sizes. The evaluation of adsorption thermodynamics showed that sulfur dioxide is adsorbed with two adsorption energies on activated carbons [101, 110], about 50 kJ/mol and 80 kJ/mol. The former represents weak physical adsorption, and the latter-chemisorption of sulfuric acid. Daley and coworkers found that the SO_2 adsorption capacity on the activated carbon fibers was inversely proportional to pore size, pore volume and pore size distribution [111].

Besides porosity, surface basicity was also found to be an important factor increasing SO_2 adsorption [101–108, 120] (Fig. 7.4). Especially, pyronic and pyronic-like groups were indicated as responsible for strong physical adsorption of sulfur dioxide [106, 108]. Oxygen-containing basic sites are proposed to be catalytic centers for the oxidation of SO_2 to SO_3 [112]. Interestingly, the oxygen present in the system can negatively influence the amount of SO_2 adsorbed [104, 106]. It is linked to its ability to react with carbonaceous matrix, its oxidation and thus a decrease in the surface area [111, 115]. On the other hand, Daley and coworkers found that for SO_2 adsorption in dry conditions, oxygen functional groups significantly enhanced the performance at temperature smaller than 348 K [111]. This effect was linked to surface reactions of quinines with SO_2 and water with the formation of diol and sulfuric acid. Raymundo-Pinero and coworkers [109] confirmed that removal of oxygen from the surface forms new high-energy adsorption/oxidation centers for SO_2. Lisovskii and coworkers [107] suggested that surface acidic groups are the catalyst for SO_2 oxidation. Rubio and Izquierdo [113] concluded that not only the amount of surface groups but also their accessibility has an effect on SO_2 adsorption/oxidation on carbonaceous materials.

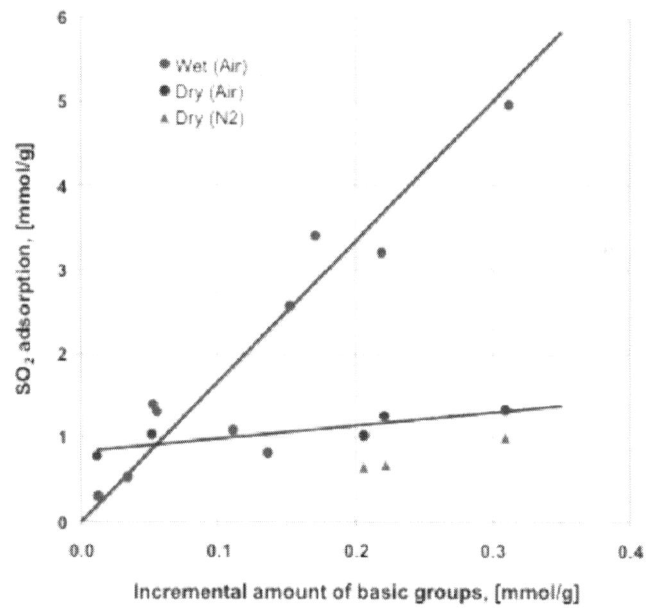

Figure 7.4 Dependence of SO_2 adsorption on the incremental basicity in pores between 6.79 and 8.58 Å. Dotted lines show the trend in the data. Reprinted with permission from Ref. [118]. Copyright American Chemical Society, 2002.

An increase in the basic nature of carbons *via* incorporation of basic nitrogen species was found to be a positive factor affecting sulfur dioxide uptake [115, 120–124]. Although the role of nitrogen present in the carbon matrix was not emphasized in the studies of PAN based activated carbon fibers used as SO_2 adsorbents, modification of their surface with pyridine and basic nitrogen functionalities increased the capacity and oxidation to sulfuric acid [124]. Raymundo-Pinero and coworkers [109] and Bagreev and coworkers [118] found that pyridinic species, which are placed at the edges of graphene layers significantly increase the amount of SO_2 adsorbed and converted to sulfuric acid, especially when the groups are located in small pores [118]. The only negative aspect of this mechanism is strong adsorption of sulfuric acid and thus the difficulties in adsorbents' regeneration [107].

The positive effect of ash on SO_2 removal and its conversion to sulfuric acid on activated carbons was indicated by Lu and Do [120] studying the SO_2 adsorption on activated coal rejected char. In their research, titanium dioxide was found to be the most active ingredient of ash. On the other hand, Bashkova and coworkers suggested CaO as the most valuable component of an inorganic mater in sewage sludge derived adsorbents [121].

Ab initio molecular orbital calculations were also performed to identify possible pathways of SO_2 oxidation on activated carbon in the presence of water and oxygen [125]. In the case surface oxides located on the zigzag edge sites, sulfuric acid was indicated as the main surface reaction products with sulfurous acid as an intermediate [125]. For edge sites containing twin oxides, chemisorption was found to be an unfavorable mechanism.

7.5 Removal of Nitric Oxides

Nitric oxides (NO_x) are formed in the combustion of fossil fuel. Here, nitrogen originates from both the N-rich fuel and from the air supplied for oxidation. Therefore, the elimination of NO_x in post-combustion approaches is important to evaluate. The most effective technique so far has been selective catalytic reduction of NO_x on various catalysts.

The adsorption of NO_x on unimpregnated activated carbons has been studied at moderate temperatures (295 K and 400 K) and at the broad range of pressure (100 and 3000 kPa) [126–130]. As much as 150 mg NO_2 per gram of carbon has been reported [126]. A pressure increase significantly enhances the NO_2 uptake, whereas an increasing temperature has an opposite effect. The micropore filling was proposed as a main adsorption mechanism. Moreover, a significant amount of NO_2 was found to be adsorbed irreversibly. Catalytic reaction of the oxidation of NO to NO_2 in the presence of oxygen with the formation of nitric acid was suggested [126].

In the case of NO reduction on chars in the temperature between 373 and 473 K, in the presence of oxygen and moisture, the capacity between 100 and 150 mg/g was recorded with reduction to N_2 (30% NO) and formation of CO_2 [127]. It was proposed that oxygen from the gas phase reacts with the carbon surface forming oxygen–carbon structures. They act as centers for NO chemisorption.

Xia and coworkers [128] found that activated carbon reduced at 1200 K in hydrogen was able to adsorb NO without its oxidation to NO_x. The amount adsorbed was about 1 mmol/g. A strong (>40 kcal/mol) and weak adsorption energies were measured. Interestingly, such a process does not take place on carbons treated in nitrogen.

Kong and Cha [129, 130] found that when carbons are treated with microwave energy of 480 W the enhanced NO_x reactions with carbon to produce nitrogen and carbon oxides take place. The microwave regeneration increased surface area and porosity. Moreover, nitrogen compounds were incorporated to the surface. It was concluded that in the presence of water and oxygen, NO is converted to NO_2 and HNO_3. A significant advantage of this process is that 90% of NO_x is reduced to N_2 during microwave treatment/regeneration/adsorption runs. An interesting aspect is the simultaneous activation of the carbon surface.

As in the case of hydrogen sulfide, activated carbons impregnated with potassium hydroxide were tested for NO_x removal [131, 132]. In these processes, KNO_2 and KNO_3 are surface reaction products. Salt crystals can block the porosity of the materials and thus decrease the efficiency for NO_x removal. Hydroxyl ions were indicated as delaying the oxidation of KNO_2 to KNO_3, which had a positive effect on NO_x removal. The effects of other potassium-containing species such as potassium carbonate or potassium silicate were also investigated for the process of NO_x adsorption on coal chars [133–136]. Coals activated at 1300 K with a high

KOH/coal ratio [133] used as NO_x adsorbents at $T < 473$ K showed the predominance of physical adsorption. On the other hand, NO_x reduction with the formation of N_2 and CO_2 was observed at $T > 473$ K. The residual potassium was indicating as a catalyst promoting the reduction process [134].

Another chemical agent promoting the reduction of NO_x is ammonia [127, 137]. Its presence on the surface of carbon fibers enhanced the process at humidity levels less than 60% [138]. As the products of surface reactions of NO with NH_3 molecular nitrogen and water were proposed [127].

Nitrogen species incorporated to the carbon matrix have also a positive effect on NO_x reduction on carbons [138]. Besides increasing the amount of NO adsorbed at room temperature, they lower the temperature at which the reaction takes place and increase the conversion ratio. The chemisorption reaction is associated with an electron transfer from the carbon surface to the NO molecule or $(NO)_2$ molecule. The resulting species were identified as diamagnetic and dimeric. Those hyponitrites are highly reactive and can easy oxidize the carbon surface resulting in the formation of N_2.

A high NO_2 removal capacity on activated carbons modified with urea [139] and dimethylamine [140] were linked to reactions of amides or amines existing on the surface with NO_2. Besides nitrates and nitrites, nitramines and nitrosoamines are the reaction products. The important aspect is that the amount of NO released decreased owing to its reaction with amines in which nitroso-amines are the products [140]. With more exposure to NO_2 the surface of the carbon matrix is reoxidized by nitrogen dioxide and NO is released. Water in the system increased the amount of NO_2 adsorbed but had a negative effect on the release of NO. The following reactions were proposed on amine-modified carbons [140]:

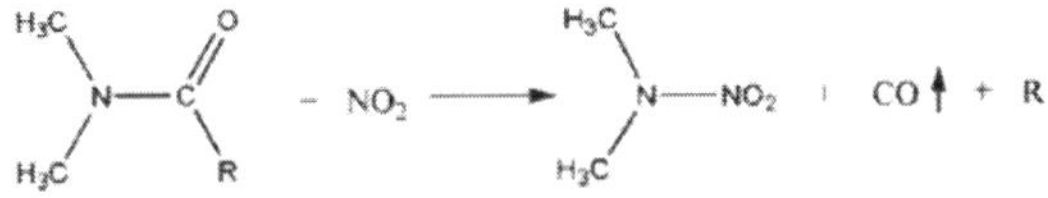

(7.14)

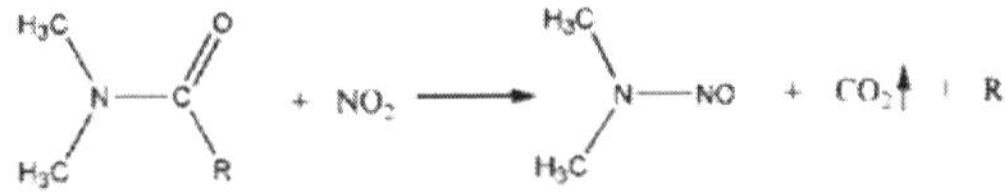

(7.15)

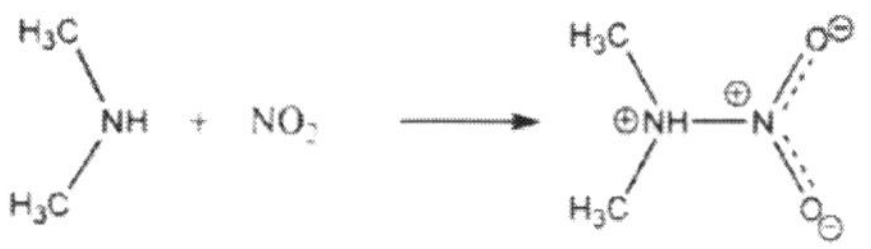

(7.16)

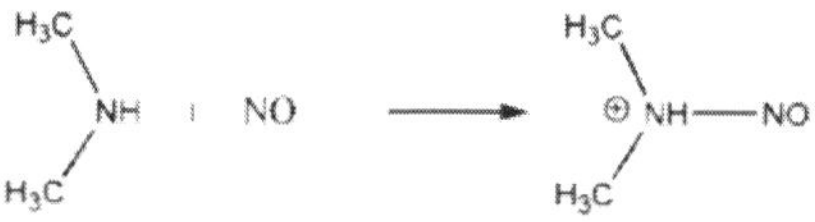

(7.17)

Besides caustic and nitrogen modifications of the carbon surfaces, the reduction of NO_x was extensively studied on carbons impregnated with transition metals such as Ni, Fe, Co, and Cu [141–145]. Copper was the most efficient catalyst in the reduction of NO to N_2 and O_2 either with or without oxygen [145]. The copper metal is active for NO removal at a high temperature even in the absence of oxygen [114].

Activated carbons with deposited copper species reduced with hydrazine or heat treatment in inert atmosphere were also studied as reactive adsorbents of NO_2. Copper loading improved the adsorption capacity of NO_2 and the retention of NO. It was linked to the presence of copper metal and its high dispersion on the surface. The results suggested that NO_2 mainly is converted to copper nitrates, although the possibility of its reduction to N_2 is not ruled out. A high capacity on hydrazine treated samples is linked to the high dispersion of metallic copper on the surface of this carbon.

Besides copper, iron was also found to have a positive effect on NO_2 adsorption activated carbons [147–149]. The results showed that the pore volume and well-dispersed iron species (Fe_2O_3, $FeSO_4$, and FeS) enhance the performance at room temperature [149]. The retention of NO_2 on the carbon surface takes place either through its reduction to NO by carbon and/or by Fe_2O_3, $FeSO_4$, and FeS, or through its reaction with Fe_2O_3 and/or $Fe(OH)_3$, leading to the formation of $Fe(NO_3)_3$. The retention of NO is enhanced on carbons containing iron in the form of α-FeOOH, α-Fe_2O_3, or γ-FeOOH. The best performance was found on the carbon with a α-FeOOH. Dispersion and the particle size of iron compounds on the carbon surface affect both the adsorption/reduction process of NO_2 and the retention of NO.

Another way of the introduction of metal species was deposition of an active inorganic phase originated from sewage sludge [150]. The results indicated that for all materials a significant amount of

NO_2 was reduced to NO and released from the system. The addition of the inorganic phase increased adsorption in moist conditions and inorganic nitrates and nitrites were formed. In dry conditions, NO_2 adsorbed and nitric and nitrous acids formed on the surface significantly increase the acidity of the carbons by the formation of oxygen-containing groups and organic nitrates.

Interesting results leading to a significant improvement in NO_2 retention on carbons at room temperature were obtained when chlorides of cerium and lanthanum were deposited in the pore system [151]. The inorganic phase, bound NO_2 in the forms of nitrates (Ce, La, Na), got oxidized/oxidized carbon surface (for Ce) or contributed to the formation of nitrosyl chloride (for Na).

When small silver particles were deposited on activated carbon surface containing phosphorus, the amount of NO_2 adsorbed at room temperature increased [152]. The products of NO_2 interactions with silver include surface chelates such as Ag_2–O–NO or Ag–O_2–NO. It was shown that both silver and phosphorus species were oxidized by NO_2 and NO was either retained on the carbon surface by its interactions with metallic silver or was further reduced to N_2O or N_2. Since the carbon support was also active in the reduction of NO_2 to NO, carbon monoxide formed in this process could reduce silver oxide nanoparticles, and thus the active silver centers were renewed.

7.6 Removal of Other Toxic Gases (HCN, AsH_3, CH_3SH)

Another gas of extremely high toxicity is hydrogen cyanide. It is an important industrial agent used in the extraction of precious metals, in electroplating industry, metallurgy, and in the production of many consumer goods. As in the case of hydrogen sulfide and sulfur dioxide, physisorption of hydrogen cyanide on activated carbons is rather weak. Therefore, chemisorption on modified carbons is a target.

Efficient systems to remove HCN contain copper and an oxidizing salt such as sodium dichromate [153]. Copper, either from a nitrate or from an oxide is reduced to metal, and then the oxidizing agent is added. The process takes place in mesopores, and cyanoformide and oxamide are the products of surface reactions. The capacity reaches about 100 mg/g at 80% humidity and 2 mg/L HCN in the challenge gas [154]. On only copper-impregnated carbon, the bulk of chemisorbed HCN is in the form of copper cyanide, and no oxamide is formed.

A quick aging process of Cu/Cr coating carbons directed the research efforts toward species forming complexes with HCN and cyanogens. On the aging-resistant carbons impregnated with nickel or cobalt acetate [154], 90 mg/g HCN was adsorbed with a mechanism similar to that on Cu-impregnated carbons. Formates and propanoates of manganese, cobalt, nickel, copper and zinc were used as impregnants of activated carbon cloth [155]. The results indicated that the reactions occur in the film of water adsorbed on the surface and their pathways follow the reactivity of the impregnated compound with HCN in the solution.

Rajakovic and coworkers [156] studied the performance of activated carbon cloth impregnated with compounds containing copper, silver (I), iron, magnesium and aluminum in the forms of acetates, oxalates, tartrates, stearates, and citrates. The results suggested that although adsorptive and chemical forces are the most important factors affecting retention of HCN, there are other important aspects influencing the performance of materials. They include the geometry of the system (porosity), diffusion rate, the kinetics of surface reactions and the type of metal salt used as an impregnant. Organic salts are more efficient than inorganic species. High concentrations of pollutants lead to the formation of complex compounds, and at low concentration the precipitation reaction is the predominant one.

Besides modifications of carbons with inorganic species, the incorporation of nitrogen to the surface as a way to enhance HCN adsorption was also investigated [157]. The nitrogen increased basicity and the ability of the surface to activate oxygen. These led to complex surface reactions in which the derivatives of hydrogen cyanide formed oxamide and were incorporated to the carbon matrix or were deposited as the bulky insoluble polymers of paracyanogen on the surface. Pores with sizes between 1 and 2 nm were found to be most active. The functional groups can be present in them, and HCN, or its derivatives, and water take part in the reactions.

The removal of arsine from a gas phase is an important issue for the semiconductor industry and also for catalytic processes, which utilize syngas. Arsine removal from syngas on various adsorbents was addressed by Quinn and coworkers [158]. As adsorbents, alumina or carbon supports with lead, manganese, copper, silver, chromium oxides, or metallic copper deposited on their surfaces were investigated. CuO showed a positive effect at high temperature (313 K). On carbon with metallic copper, besides Cu_3As, hydrogen was also produced. Although it was suggested that oxygen groups of the carbon surface are active sites for AsH_3 binding [158], a detailed role of those centers was not addressed. In the 1940s, Hickey and Wiig also found that on the carbons containing copper oxides, arsine was oxidized to arsenic trioxide [159]. On the carbon with cupric oxide, about 130 mg of AsH_3 per cm^3 of adsorbent (or about 270 mg/g) was adsorbed at dry conditions. Although the role of water was mentioned, it was not addressed in detail.

A recent study on the removal of arsine on carbon modified with sulfur [160] or nitrogen [161] demonstrated that oxygen- and sulfur-containing groups participate in arsine oxidation to arsenic tri- and pentoxide and/or in the formation of arsenic sulfides. This occurred either *via* the activation of oxygen on these groups and on nitrogen-containing groups [161] or *via* a direct involvement of sulfur groups in the reactions with arsine. The hydrophilic surface of carbon enhances the removal of arsine *via* retention of small quantity of water from the environment. At high humidity levels, water fills the pore system and blocks the catalytic action of the surface toward oxidation and carbon exhibit nil capacity.

To remove methyl mercaptan (MM) from the air, activated carbons are also used [162–168]. On their surface, methyl mercaptan is oxidized to dimethyl disulfide (DMDS), which is much strongly adsorbed at room temperature than methyl mercaptan. This happens owing to the larger size of DMDS molecule and higher boiling point than those for MM [169]. Activated carbons impregnated with chemicals such as potassium iodide, potassium iodate, potassium carbonate, and ammonia have also been reported as good adsorbents of methyl mercaptan adsorbents [165]. Surface oxygen or nitrogen groups and porosity were also indicated as features governing MM removal on virgin carbons [170]. The results showed that the main product of oxidation, dimethyl disulfide, is adsorbed in pores smaller than 5 nm [171]. It competes with water for adsorption sites. Nevertheless, dimethyl disulfide wins owing to its strong adsorption in the carbon pore system. It is interesting that dimethyl disulfide cannot be formed in a significant quantity without water deposited on the surface. When the carbon surface pH is above the threshold of about 9.5, water facilitates the dissociation of methyl mercaptan, and in this form it is oxidized to DMDS [171]. Besides surface oxygen and nitrogen basic groups, iron oxide also contributes to MM oxidation *via* redox reactions [170].

7.7 Concluding Remarks

The literature dedicated to the removal of toxic gases is extensive. This chapter mainly focuses on the removal of selected gases at ambient temperatures. At these conditions, only separations based on reactive adsorption or chemisorption can provide sufficient efficiencies. Table 7.1 summarizes important surface features needed in carbon-based adsorbents. It also provides the most important surface transformations identified so far through extensive studies performed by various research groups. Although new materials are under constant development, and the removal of toxic gases is one of the important issues facing the contemporary society, activated carbons, especially those with modified/designed surface features, still outperform the new materials such as metallic organic frameworks in terms of combined costs, availability, and removal capacity. Figure 7.5 presents the comparison of the examples of the breakthrough capacities measured on activated carbons with those reported so far on MOF [172–174]. As it can be seen, activated carbons are still an important technology for air purification in the future.

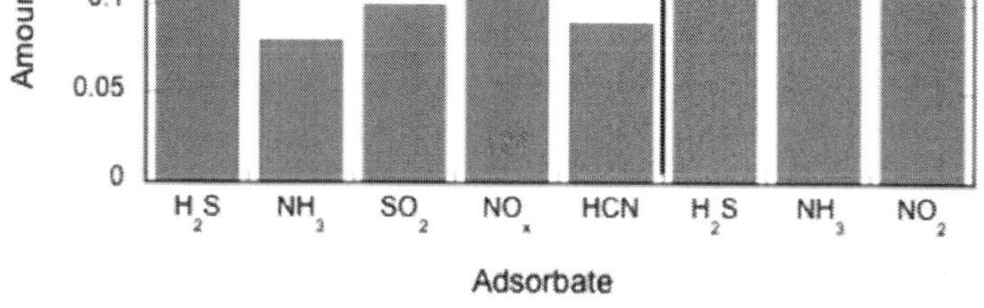

Figure 7.5 Comparison of the toxic gases removal capacities at ambient conditions on activated carbons to those on copper-based metal organic frameworks (MOFs).

Table 7.1 Summary of important surface features governing the adsorption of inorganic species and VOCs at the temperatures close to ambient in the presence of air and humidity

Adsorbate	Mechanism	Surface reaction products	Role of humidity	Active texture feature	Active surface chemistry	Ash/ inorganic matter
NH_3	Dissolution. Complexation/ oxidation/ acid–base reactions	Ammonium hydroxide, ammonium salts of carboxylic aids/ amines/amides	Enhanced, participate reactions and dissolution	Micropores < 1 nm to attract NH_3	Acidic groups, metals salts	—
H_2S	Physical adsorption/ dissolution/ oxidation	Sulfur radicals, Sulfur polymers, SO_2, H_2SO_4	Crucial component; Adsorption in film of water, ensures dissociation	Volume of micropores Small size (<1 nm) promotes formation of SO_2 and H_2SO_4	Basic surface (nitrogen or oxygen groups) promotes dissociation of H_2S and oxidation of HS^-, required pH > 4.5; basic nitrogen groups significantly increase basicity	Fe_2O_3, CaO, MgO, caustics, $NaHCO_3$ Cu, Metals from group 6–8
SO_2	Physical adsorption/ oxidation	SO_2, H_2SO_4	Enhanced removal	Pore width < 0.7 nm Volume of micropores	Active centers—oxygen- and nitrogen-containing groups	CaO

NO_x	Adsorption/ oxidation/ reduction	NO, NO_2, N_2, CO_2, CO, HNO_3 nitrites, nitramines, nitrosoamines	Should be less than 60%		Oxygen- or nitrogen-containing groups are centers for NO chemisorption	Cu, Ce, La, Ag
HCN	Adsorption/ complexation/ oxidation	$(CN)_2$, CN-$CONH_2$, $(CONH_2)_2$ metal cyanides	Enhanced removal	Mesopores where impregnant is present	Nitrogen groups	Cu, Ag, Fe, Ni, Zn, Cr
AsH_3	oxidation	AsO_3, As_2O_5, arsenic acids	A small amount is beneficial.	High volume of micropores	Oxygen, nitrogen or sulfur groups activating oxygen	Cu
CH_3SH	Physical adsorption/ oxidation	Dimethyldisulufide	Should be less than 5 nm in diameter	A high volume of pores less than 5 nm is needed	Oxygen and nitrogen groups important for enhancement of oxidation, pH > 9.5 important for dissociation	Fe

References

1. Manahan, S. E. (1997). *Environmental Chemistry*, 7th ed., CRC Press.
2. http://www.epa.gov/air/airpollutants.html.
3. http://www.epa.gov/ttn/atw/allabout.html.
4. Helminem J., Helenius, J., and Paatero, E. (2001). Adsorption equilibria of ammonia gas on inorganic and organic sorbents at 298.15 K. *J. Chem. Eng. Data*, **46**, 391–399.
5. Stoeckli, F., Guillot, A., and Slasli, M. (2004). Specific and non-specific interactions between ammonia and activated carbons. *Carbon*, **42**, 1619–1624.
6. Le Leuch, L. M., and Bandosz, T. J. (2007). The role of water and surface acidity on the reactive adsorption of ammonia on modified activated carbon. *Carbon*, **45**, 568–578.
7. Bandosz, T. J., and Ania, C. O. (2006). Surface chemistry of activated carbons and its characterization, in *Activated Carbon Surfaces in Environmental Remediation* (Bandosz, T. J., ed.), Elsevier, Oxford, 159–230.
8. Bandosz, T. J. (2006). Desulfurization on activated carbons, in *Activated Carbon Surfaces in Environmental Remediation* (Bandosz, T. J., ed.), Elsevier, Oxford, 231–292.
9. Bottani, E. J., and. Tascon, J. M. D. (eds.) (2005). *Adsorption by Carbons*, Elsevier, Amsterdam.
10. Bansal, R. C., and Goyal, M. (2005). *Activated Carbon Adsorption*, CRC Press.
11. Thompson, J. C. (1971). Compressibility of metal-ammonia solutions. *Phys. Rev. A*, **4**, 802–804.
12. Leon y Leon, C. A., and Radovic, L. R. (1992). Interfacial chemistry and electrochemistry of carbon surfaces, in *Chemistry and Physics of Carbon* (Thrower, P. A., and Dekker, M., eds.), CRC Press. Vol. 24, pp. 213–310.
13. Mangun, C. L., Braatz, R. D., Economy, J., and Hall, A. J. (1999). Fixed bed adsorption of acetone and ammonia onto oxidized activated carbon fibers. *Ind. Eng. Chem. Res.*, **38**, 3499–3504.
14. Tamon, H., and Okazaki, M. (1996). Influence of acidic surface oxides of activated carbon on gas adsorption characteristics. *Carbon*, **34**, 741–746.

15. Park, S.-J., and Kim, B.-J. (2005). Ammonia removal of activated carbon fibers produced by oxyfluorination. *J. Colloid Interface Sci.*, **291**, 597–599.

16. Mangun, C. L., Benak, K. R., Daley, M. A., and Economy, J. (1999). Oxidation of activated carbon fibers: effect on pore size, surface chemistry, and adsorption properties. *Chem. Mater.*, **11**, 3476–3483.

17. Park, S.-J., and Kim, K. D. (1999). Adsorption behaviors of CO_2 and NH_3 on chemically surface-treated activated carbons. *J. Colloid Interface Sci.*, **212**, 186–189.

18. Petit, C., and Bandosz, T. J. (2009). Role of surface heterogeneity in the removal of ammonia from air on micro/mesoporous activated carbons modified with molybdenum and tungsten oxides. *Micro. Meso. Mater.*, **118**, 61–67.

19. Seredych, M., and Bandosz, T. J. (2010). Combined role of water and surface chemistry in reactive adsorption of ammonia on graphite oxides. *Langmuir*, **26**, 5491–5498.

20. Seredych, M., Tamashausky, A. V., and Bandosz, T. J. (2010). Graphite oxides obtained from porous graphite: the role of surface chemistry and texture in ammonia retention at ambient conditions. *Adv. Functional. Mater.*, **20**, 1670–1679.

21. Petit, C., Karwacki, C., Peterson, G., and Bandosz, T. J. (2007). Interactions of ammonia with the surface of microporous carbon impregnated with transition metal chlorides. *J. Phys. Chem. C*, **111**, 12705–12714.

22. Petit, C., and Bandosz, T. J. (2009). Graphite oxide/polyoxometalate nanocomposites as adsorbents of ammonia. *J. Phys. Chem. C*, **113**, 3800–3809.

23. Seredych, M., and Bandosz, T. J. (2009). Graphite oxide/AlZr polycation composites: surface characterization and performance as adsorbents of ammonia. *Mater. Chem. Phys.*, **117**, 99–106.

24. Huang, C.-C., Li, H.-S., and Chen, C.-H. (2008). Effects of surface acidic oxides of activated carbon on adsorption of ammonia. *J. Hazard. Mater.*, **159**, 523–527.

25. Kim, B.-J., and Park, S.-J. (2007). Effect of carbonyl groups formation on ammonia adsorption of porous carbon surfaces. *J. Colloid Interface Sci.*, **311**, 311–314.

26. Park, S.-J., and Jin, S.-Y. (2005). Effects of ozone treatment on ammonia removal of activated carbons. *J. Colloid Interface Sci.*, **286**, 417–419.

27. Petit, C., Seredych, M., and Bandosz, T. J. (2009). Revisiting the chemistry of graphite oxides and its effect on ammonia adsorption. *J. Mater. Chem.*, **19**, 9077–9185.

28. Petit, C., Kante, K., and Bandosz, T. J. (2010). The role of sulfur-containing groups in ammonia retention on activated carbons. *Carbon*, **48**, 654–667.

29. Bandosz, T. J., and Petit, C. (2009). On the reactive adsorption of ammonia on activated carbons modified by impregnation with inorganic compounds. *J. Colloid Interface Sci.*, **338**, 329–345.

30. Seredych, M., Rossin, J. A., and Bandosz, T. J. (2011). Changes in graphite oxide texture and chemistry upon oxidation and reduction and their effect on adsorption of ammonia. *Carbon*, **42**, 4392–4402.

31. Smith, J. W. H., Westreich, P., Smith, A. J., Fortier, H., Croll, L. M., Reynolds, J. H., and Dahn, J. R. (2010). Investigation of copper oxide impregnants prepared from various precursors for respirator carbons. *J. Colloid Interface Sci.*, **241**, 162–170.

32. Fortier, H., Westreich, P., Zelenietz, C., and Danh, J. R. (2008). Ammonia, cyclohexane, nitrogen and water adsorption capacities of an activated carbon impregnated with increasing amounts of $ZnCl_2$, and designed to chemisorb gaseous NH_3 from air stream. *J. Colloid Interface Sci.*, **320**, 423–435.

33. Molina-Sabio, M., Gonzalez, J. C., and Rodriguez-Reinoso, F. (2004). Adsorption of NH_3 and H_2S on activated carbon and activated carbon-sepiolite pellets. *Carbon*, **42**, 423–460.

34. Turk, A., Sakalis, S., Lessuck, J., Karamitsos, H., and Rago, O. (1989). Ammonia injection enhances capacity of activated carbon for hydrogen sulfide and methyl mercaptan. *Environ. Sci. Technol.*, **33**, 1242–1245.

35. Stuetz, R. M., Fenner, R. M., and Engin, G. (1999). Assessment of odours from sewage treatment works by an electronic nose, H_2S analysis and olfactometry. *Water Res.*, **33**, 453–461.

36. Turk, A., Mahmood, K., and Mozaffari, J. (1993). Activated carbon for air purification in New York City's sewage Treatment plants. *Water Sci. Technol.*, **27**, 121–126.

37. Turk, A., and Bandosz, T. J. (2000). Adsorption systems for odour treatment, in *Odours in Wastewater Treatment: Measurement, Modeling and Control* (Stuetz, R. M., and Frechen, F.-B. eds.), IWA, pp. 354–364.

38. Bandosz, T. J., Bagreev, A., Adib, F., and Turk, A. (2000). Unmodified versus caustics-impregnated carbons for control of hydrogen sulfide

emissions from sewage treatment plants. *Environ. Sci. Technol.*, **34**, 1069–1074.

39. Yan, R., Liang, D. T., Tsen, L., and Tay J. H. (2002). Kinetics and mechanisms of H_2S adsorption by alkaline activated carbon. *Environ. Sci. Technol.*, **36**, 4460–4466.

40. Chiang H.-L., Tsai, J.-H., Tsai C.-L., and Hsu, Y.-C. (2000). Adsorption characteristics of alkaline activated carbon exemplified by water vapor, H_2S and CH_3SH gas. *Sep. Sci. Technol.*, **35**, 903–918.

41. Bagreev, A., and Bandosz, T. J. (2002). A role of sodium hydroxide in the process of hydrogen sulfide adsorption/oxidation on caustic-impregnated activated carbons. *Ind. Eng. Chem. Res.*, **41**, 672–679.

42. ASTM Standards. Vol. 15.01. *Refractories; Carbon and Graphite Products; Activated Carbon; Advanced Ceramics*. ASTM D6646-01. 1998.

43. Bandosz, T. J., and Le, Q. (1998). Evaluation of surface properties of exhausted carbons used as H_2S adsorbents in sewage treatment plants. *Carbon*, **36**, 39–44.

44. Przepiorski, J., and Oya, A. (1998). K_2CO_3-loaded deodorizing activated carbon fibre against H_2S gas: factors influencing the deodorizing efficiency and the regeneration method. *J. Mater. Sci. Lett.*, **17**, 679–682.

45. Przepiorski, J., Yoshida, and S., Oya, A. (1999). Structure of K_2CO_3-loaded activated carbon fiber and its deodorization ability against H_2S gas. *Carbon*, **37**, 1881–1890.

46. Przepiorski, J., Abe, Y., Yoshida, S., and Oya, A. (1997). Preferential supporting of potassium carbonate around the peripheral region of activated carbon fibre. *J. Mater. Sci. Lett.*, **16**, 1312–1314.

47. Coskun, I., and Tollefson, E. L. (1986). Oxidation of low concentrations of hydrogen sulfide over activated carbons. *Can. J. Chem. Eng.*, **58**, 72–76.

48. Steijns, M., and Mars, P. (1977). Catalytic oxidation of hydrogen sulfide. Influence of pore structure and chemical composition of various porous substances. *Ind. Eng. Chem. Prod. Res. Dev.*, **16**, 35–41.

49. Ghosh, T. K., and Tollefson, E. L. (1986). A continuous process for recovery of sulfur from natural gas containing low concentrations of hydrogen sulfide. *Can. J. Chem. Eng.*, **64**, 960–968.

50. Ghosh, T. K., and Tollefson, E. L. (1986). Kinetics and reaction mechanism of hydrogen sulfide oxidation over activated carbon in the temperature range of 125–200°C. *Can. J. Chem. Eng.*, **64**, 969–976.

51. Dalai, A. K., Majumadar, M., Chowdhury, A., and Tollefson, E. L. (1993). The effects of pressure and temperature on the catalytic oxidation of hydrogen sulfide in natural gas and regeneration of the catalysts to recover the sulfur produced. *Can. J. Chem. Eng.*, **71**, 75–82.

52. Yang, A., Tollefson, E. L., and Dalai, A. K. (1998). Oxidation of low concentrations of hydrogen sulphide: process optimization and kinetics studies. *Can. J. Chem. Eng.*, **76**, 76–86.

53. Dalai, A. K., and Tollefson, E. L. (1986). Kinetics and reaction mechanism of catalytic oxidation of low concentrations of hydrogen sulfide in natural gas over activated carbon. *Can. J. Chem. Eng.*, **76**, 902–914.

54. Dalai, A. K., Majumdar, A., and Tollefson, E. L. (1999). Low temperature catalytic oxidation of hydrogen sulfide in sour produced waste-water using activated carbon catalysts. *Environ. Sci. Technol.*, **33**, 2241–2246.

55. Meeyoo, V., Trimm, D. L., and Cant, N. W. (1997). Adsorption-reaction Processes for the removal of hydrogen sulphide from gas streams. *J. Chem. Technol. Biotechnol.*, **68**, 411–416.

56. Steijns, M., Derks, F., Verloop, A., and Mars, P. (1976). The mechanism of the catalytic oxidation of hydrogen sulfide. II. Kinetics and mechanism of hydrogen sulfide oxidation catalyzed by sulfur. *J. Catal.*, **42**, 87–95.

57. Steijns, M., Koopman, P., Nieuwenhuijse, B., and Mars, P. (1976). The mechanism of the catalytic oxidation of hydrogen sulfide. III. An electron spin resonance study of the sulfur catalyzed oxidation of hydrogen sulfide. *J. Catal.*, **42**, 96–106.

58. Steijns, M., and Mars, P. (1974). The role of sulfur trapped in micropores in the catalytic partial oxidation of hydrogen sulfide with oxygen. *J. Catal.*, **35**, 11–17.

59. Klein, J., and Henning, K.-D. (1984). Catalytic oxidation of hydrogen sulphide on activated carbons. *Fuel*, **63**, 1064–1067.

60. Hedden, K., Humber, L., and Rao, B. R. (1976). *Adsorptive Reinigung von schwefelwasserstoffhaltigen abgasen*. VDI-Bericht Nr. 253 S. 37/42.

61. Katoh, H., Kuniyoshi, I., Hirai, M., and Shoda, M. (1995). Studies of the oxidation mechanism of sulphur-containing gases on wet activated carbon fibre. *Appl. Catal. B: Environ.*, **6**, 255–262.

62. Tanada, S., Kita, T., Boki, K., and Kozaki, Y. (1985). Preparation of narrow pores carbon suitable for hydrogen sulfide adsorption. *J. Environ. Sci. Health A*, **20**, 87–96.

63. Choi, J. J., Hirai, M., and Shoda, M. (1991). Catalytic oxidation of hydrogen sulphide by air over an activated carbon fibre. *App. Catal. A*, **7**, 241–248.

64. Sreeramamurthy, R., and Menon, P. G. (1975). Oxidation of H_2S on active carbon catalysts. *J. Catal.*, **37**, 287–296.

65. Primavera, A., Trovarelli, A., Andreussi, P., and Dolcetti, G. (1998). The effect of water in the low-temperature catalytic oxidation of hydrogen sulfide to sulfur over activated carbon. *Appl. Catal. A: Gen.*, **173**, 185–192.

66. Kaliva, A. N., and Smith, J. W. (1983). Oxidation of low concentrations of hydrogen sulfide by air on a fixed activated carbon bed. *Can. J. Chem. Eng.*, **61**, 208–212.

67. Le Lauch, L. M., Subrenat, A., and Le Cloirec, P. (2003). Hydrogen sulfide adsorption and oxidation onto activated carbon cloth: applications to odorous gaseous emission treatments. *Langmuir*, **9**, 10869–10877.

68. Mikhalovsky, S. V., Zaitsev, and Yu. P. (1997). Catalytic properties of activated carbons I. Gas-Phase oxidation of hydrogen sulphide. *Carbon*, **35**, 1367–1374.

69. Bandosz, T. J. (1999). Effect of pore structure and surface chemistry of virgin activated carbon on removal of hydrogen sulfide. *Carbon*, **37**, 483–491.

70. Adib, F., Bagreev, A., and Bandosz, T. J. (2000). Analysis of the relationship between H_2S removal capacity and surface properties of unmodified activated carbons. *Environ. Sci. Technol.*, **34**, 686–692.

71. Adib, F., Bagreev, A., and Bandosz, T. J. (1999). Effect of surface characteristics of wood based activated carbons on removal of hydrogen sulfide. *J. Colloid Interface Sci.*, **214**, 407–415.

72. Adib, F., Bagreev, A., and Bandosz, T. J. (1999). Effect of pH and surface chemistry on the mechanism of H_2S removal by activated carbons. *J. Colloid Interface Sci.*, **216**, 360–369.

73. Adib, F., Bagreev, A., and Bandosz, T. J. (2000). On the possibility of regeneration of unimpregnated activated carbons used as hydrogen sulfide adsorbents. *Ind. Eng. Chem. Res.*, **39**, 2439–2446.

74. Bagreev, A., Rahman, H., and Bandosz, T. J. (2000). Wood-based activated carbons as adsorbents of hydrogen sulfide: a study of adsorption and water regeneration process. *Ind. Eng. Chem. Res.*, **39**, 3849–3855.

75. Bagreev, A., Rahman, H., and Bandosz, T. J. (2000). Study of H_2S adsorption and water regeneration of coconut-based activated carbon. *Environ. Sci. Technol.*, **34**, 4587–4592.

76. Bagreev, A., Adib, F., and Bandosz, T. J. (1999). Initial heats of H_2S adsorption on activated carbons: effect of surface features. *J. Colloid Interface Sci.*, **219**, 327–332.

77. Bagreev, A., Adib, F., and Bandosz, T. J. (2001). pH of the activated carbon surface as an indication for its suitability for removal of hydrogen sulfide from wet air streams. *Carbon*, **39**, 1987–1905.

78. Bagreev, A., and Bandosz, T. J. (2001). H_2S adsorption/oxidation on unmodified activated carbons: importance of prehumidification. *Carbon*, **39**, 2303–2311.

79. Hayden, R. A. (1995). Process for making catalytic carbon. *U.S. Patent*, 5,444,031.

80. Bagreev. A., Menendez, J. A., Dukhno, I., Tarasenko, Y., and Bandosz, T. J. (2004). Bituminous coal-based activated carbons modified with nitrogen as adsorbents of hydrogen sulfide. *Carbon*, **42**, 469–476.

81. Hayden, R. A. (1995). Process for regenerating nitrogen-treated carbonaceous chars used for hydrogen sulfide removal. *International Patent WO* 95/26230.

82. Matviya, T. M., and Hayden, R. A. (1994). Catalytic carbon. *U.S. Patent*, 5,356,849.

83. Turk, A., Sakalis, E., Rago, O., and Karamitsos, H. (1992). Activated carbon systems for removal of light gases. *Ann. N. Y. Acad. Sci.*, **661**, 221–228.

84. Boudou, J. P., Chehimi, M., Broniek, E., Siemieniewska, T., and Bimer, J. (2003). Adsorption of H_2S or SO_2 on an activated carbon cloth modified by ammonia treatment. *Carbon*, **41**, 1999–2007.

85. Yang, Q. H., Zheng, J. T., Li, Y., Wang, M. Z., and Zhang, B. J. (1999). Adsorption and conversion of hydrogen sulfide over PAN-based ACF. *Carbon*, **37**, 2078–2080.

86. Adib, F., Bagreev, A., and Bandosz, T. J. (2000). Adsorption/oxidation of hydrogen sulfide on nitrogen modified activated carbons. *Langmuir*, **16**, 1980–1986.

87. Kastner, J. R., Das, K. C., and Melear, N. D. (2002). Catalytic oxidation of gaseous reduced sulfur compounds using coal fly ash. *J. Hazard. Mater. B*, **95**, 81–90.

88. Kastner, J. R., Das, K. C., Buquoi, Q., and Melear, N. D. (2003). Low temperature catalytic oxidation of hydrogen sulfide and methanethiol using wood and coal fly ash. *Environ. Sci. Technol.*, **37**, 2568–2574.

89. Cal, M. P., Strickler, B. W., and Lizzio, A. A. (2000). High temperature hydrogen sulfide adsorption on activated carbon. I. Effect of gas composition and metal addition. *Carbon*, **38**, 1757–1765.

90. Bansal, R. C., Donnet, J. B., and Stoeckli, F. (1988). *Active Carbon*, Marcel Dekker.

91. Salame, I. I., and Bandosz, T. J. (1999). Study of water adsorption on activated carbons with different degrees of surface oxidation. *J. Colloid Interface Sci.*, **210**, 367–374.

92. Salame, I. I., and Bandosz, T. J. (1999). Revisiting the effect of surface chemistry on adsorption of water on activated carbons. *J. Phys. Chem.*, **103**, 3877–3884.

93. Boulinguiez B., and, Le Cloirec, P. (2010). Chemical transformations of sulfur compounds adsorbed onto activated carbon materials during thermal desorption. *Carbon*, **48**, 1558–1569.

94. Bagreev, A., Bashkova, S., Locke, D. C., and Bandosz, T. J. (2001). Sewage sludge derived materials as efficient adsorbents for removal of hydrogen sulfide. *Environ. Sci. Technol.*, **35**, 1537–1543.

95. Bagreev, A., and Bandosz, T. J. (2005). On the mechanism of hydrogen sulfide removal from moist air on catalytic carbonaceous adsorbents. *Ind. Chem. Eng. Res.*, **44**, 530–539.

96. Seredych, M., and Bandosz, T. J. (2007). Desulfurization of digester gas on industrial sludge-derived adsorbents. *Energy Fuels*, **21**, 858–866.

97. Seredych, M., and Bandosz, T. J. (2006). Desulfurization of digester gas on catalytic carbonaceous adsorbents: complexity of interactions between the surface and components of the gaseous mixture. *Ind. Chem. Eng. Res.*, **45**, 3658–3665.

98. Bandosz, T. J. (2002). On the adsorption/oxidation of hydrogen sulfide on unmodified activated carbon at ambient conditions. *J. Colloid Interface Sci.*, **246**, 1–20.

99. Kohl, A., Riesenfeld, F. (1985). *Gas Purification*, 4th ed., Gulf Publishing Company.

100. Stirling, D. (2000). *The Sulfur Problem: Cleaning Up Industrial Feedstocks*, The Royal Society of Chemistry.

101. Rodriguez-Mirasol, J., Cordero, T., Rodriguez, J. J. (1997). Effect of oxygen on the adsorption of SO_2 on activated carbon. Abstract of *23rd Biennial Conference on Carbon*, 18–23 July, College Park, PA, p. 376.

102. Moreno-Castilla, C., Carrasco-Marin, F., Utrera-Hidalgo, E., and Rivera-Utrilla, J. (1993). Activated carbons as adsorbents of sulfur dioxide in flowing air. Effect of their pore texture and surface basicity. *Langmuir*, **9**, 1378–1383.

103. Lisovskii, A., Shter, G. E., Semiat, R., and Aharoni, C. (1997). Adsorption of sulfur dioxide by active carbon treated by nitric acid: II. Effect of preheating on the adsorption properties. *Carbon*, **35**, 1645–1648.

104. Davini, P. (2001). SO_2 adsorption by activated carbons with various burnoffs obtained from bituminous coal. *Carbon*, **39**, 1387–1393.

105. Mochida, I., Miyamoto, S., Kuroda, K., Kawano, S., Yatsunami, S., Korai, Y., Yatsutake, A., and Yashikawa, M. (1999). Adsorption and adsorbed species of SO_2 during its oxidative removal over pitch-based activated carbon fibers. *Energy Fuels*, **13**, 369–373.

106. Davini, P. (1990). Adsorption and desorption of SO_2 on active carbon: the effect of surface basic groups. *Carbon*, **28**, 565–571.

107. Lisovskii, A., Semiat, R., and Aharoni, C. (1997). Adsorption of sulfur dioxide by active carbon treated by nitric acid: I. Effect of the treatment on adsorption of SO_2 and Extractability of the acid formed. *Carbon*, **35**, 1639–1643.

108. Anurov, C. A. (1996). Physicochemical aspects of the adsorption of sulfur dioxide by carbon adsorbents. (*Uspekhi Khimii*) *Russ. Chem. Rev.*, **65**, 663–676.

109. Raymundo-Piñero, E., Cazola-Amorós, D., Salinas-Martinez de Lecea, C., and Linares-Solano, A. (2000). Factors controlling the SO_2 removal by porous carbons: relevance of the SO_2 oxidation steep. *Carbon*, **38**, 335–344.

110. Molina-Sabio, M., Muñecas, M. A., Rodriguez-Reinoso, F., McEnaney, B (1995). Adsorption of CO_2 and SO_2 on activated carbons with a wide range of micropore size distribution. *Carbon*, **33**, 1777–1782.

111. Daley, M. A., Mangun, C. L., DeBarr, J. A., Riha, S., Lizzio, A. A., Donnals, G. L., and Economy, J. (1997). Adsorption of SO_2 onto oxidized and heat-treated activated carbon fibers (ACFS). *Carbon*, **35**, 411–417.

112. Lizzio, A. A., and DeBarr, J. A. (1997). Mechanism of SO_2 removal by carbon. *Energy Fuels*, **11**, 284–291.

113. Rubio, B., and Izquierdo, M. T. (1997). Influence of low-rank coal char properties on their SO_2 removal capacity from flue gases: I. non-activated chars. *Carbon*, **35**, 100–1011.

114. Rubio, B., Izquierdo, M. T., and Mastral, A. M. (1998). Influence of low-rank coal char properties on their SO_2 removal capacity from flue gases. 2. Activated chars. *Carbon*, **36**, 263–268.

115. Mochida, I., Korai, Y., Shirahama, M., Kawano, S., Hada, T., Seo, Y., Yoshikawa, M., and Yasutake, A. (2000). Removal of SO_x and NO_x over activated carbon fibers. *Carbon*, **38**, 22–239.

116. Davini, P., and Stoppato, G. (1997). SO_2 adsorption on active carbons: the effect of certain metal compounds. Abstract of *23rd Biennial Conference on Carbon*, July 18–23, College Park, PA, p. 316.

117. Roman, M. C., Takarada, T., Suzuki, Y., and Linares, A. (1997). SO_2 interaction with a Ca-exchanged-coal. Abstract of *23rd Biennial Conference on Carbon*, July 18–23, College Park, PA, p. 324.

118. Bagreev, A., Bashkova, S., and Bandosz, T. J. (2002). Adsorption of SO_2 on activated carbons: the effect of nitrogen functionality and pore sizes. *Langmuir*, **18**, 1257–1264.

119. Wang, Z.-M., and Kaneko, K. (1998). Effect of pore width on micropore filling mechanism of SO_2 in carbon micropores. *J. Phys. Chem. B*, **102**, 2863–2868.

120. Lu, G. Q., and Do, D. D. (1993). Retention of sulfur dioxide as sulfuric acid by activated coal reject chart. *Sep. Technol.*, **3**, 106–110.

121. Bashkova, S., Bagreev, A., Locke, D. C., and Bandosz, T. J. (2001). Adsorption of SO_2 on sewage sludge-derived materials. *Environ. Sci. Technol.*, **35**, 3263–3269.

122. Ho, C.-S., and Shih, S.-M. (1992). $Ca(OH)_2$/fly ash sorbents for SO_2 removal. *Ind. Eng. Chem. Res.*, **31**, 1130–1135.

123. Kawabuchi, Y., Sotowa, C., Kuroda, K., Kawano, S., Whitehurst, D., and Mochida, I. (1996). Preparation of active carbon fiber with basic properties. Abstracts. *Int. Conf. Carbon*, Carbon'96, Newcastle. UK, p. 431.

124. Lee, J. K., Shim, H. J., Lim, J. C., Choi, G. J., Kim, Y. D., Min, B. G., and Park, D. (1997). Influence of tension during oxidative stabilization on SO_2 adsorption characteristics of polyacrylonitrile (PAN) based activated carbon fibers. *Carbon*, **35**, 837–843.

125. Yang, F. H., and Yang, R. T. (2003). An initio molecular orbital study of the mechanism of SO_2 oxidation catalyzed by carbon. *Carbon*, **41**, 2149–2158.

126. Rubel, A. M., and Stencel, J. M. (1996). Effect of Pressure on NO_x adsorption by activated carbons. *Energy Fuels*, **10**, 704–70.

127. Izquierdo, M. T., and Rubio, B. (1998). Influence of char physicochemical features on the flue gas nitric oxide reduction with chars. *Environ. Sci. Technol.*, **32**, 4017–4022.

128. Xia, B., Phillips, J., and Chen, C.-K. (1999). Impact of pretreatments on the selectivity of carbon for NO_x adsorption/reduction. *Energy Fuels*, **13**, 903–906.

129. Kong, Y., and Cha, C.-Y. (1996). NO_x abatement with carbon adsorbents and microwave energy. *Energy Fuels*, **9**, 971–975.

130. Kong, Y., and Cha, C.-Y. (1996). Microwave-induced regeneration of NO_x-saturated char. *Energy Fuels*, **10**, 1245–1249.

131. Lee, Y.-W., Park, J.-W., Yun, J.-H., Lee, J.-H., and Choi, D.-K. (2002). Studies on the surface chemistry based on competitive adsorption of NO_x-SO_2 onto a KOH impregnated activated carbon in excess O_2. *Environ. Sci. Technol.*, **36**, 4928–4935.

132. Lee, Y.-W., Choi, D.-K., and Park, J.-W. (2001). Surface chemical characterization using AES/SAM and ToF-SIMS on KOH-impregnated activated carbon by selective adsorption of NO_x. *Ind. Eng. Chem. Res.*, **40**, 3337–3345.

133. Illan-Gomez, M. J., Salina-Martinez de Lecea, C., Linares-Solano, A., and Radovic, L. R. (1998). Potassium-containing coal chars as catalysts for NO_x reduction in the presence of oxygen. *Energy Fuels*, **12**, 1256–1264.

134. Garcia-Garcia, A., Illan-Gomez, M. J., Linares-Solano, A., and Salinas-Martinez de Lecea, C. (2002). NO_x reduction by potassium-containing coal briquettes. Effect of preparation procedure and potassium content. *Energy Fuels*, **16**, 569–574.

135. Garcia-Garcia, A., Illan-Gomez, M. J., Linares-Solano, A., and Salinas-Martinez de Lecea, C. (1999). NO_x reduction by potassium-containing coal briquettes. Effect of NO_2 concentration. *Energy Fuels*, **13**, 499–505.

136. Bueno-Lopez, A., Caballero, J. A., and Garcia-Garcia, A. (2002). Analysis of the reaction conditions in the NO_x reduction process by carbon

with a view to achieve high NO_x conversions. Residence time considerations. *Energy Fuels*, **16**, 1425–1428.

137. Mochida, I., Korai, Y., Shirahama, M., Kawano, S., Hada, T., Seo, Y., Yoshokawa, M., and Yasutake, A. (2000). Removal of SO_x and NO_x over activated carbon fibers. *Carbon*, **38**, 227–239.

138. Matzer, S., and Boehm, H. P. (1998). Influence of nitrogen doping on the adsorption and reduction of nitric oxide by activated carbons. *Carbon*, **36**, 1697–1709.

139. Oark, B.-J., Park, S.-J., and Ryu, S.-K. (1999). Removal of NO over copper supported on activated carbon prepared by electroless plating. *J. Colloid Interface Sci.*, **217**, 142–145.

140. Bashkova, S., and Bandosz, T. J. The effects of urea modification and heat treatment on the process of NO_2 removal by wood-based activated carbon. *J. Colloid Interface Sci.*, **330**, 60–66.

141. Deliyanni, E., and Bandosz, T. J. (2011). Effect of carbon surface modification with dimethylamine on reactive adsorption of NO_x. *Langmuir*, **27**, 1837–1843.

142. Davini, P. (2001). SO_2 and NO_x adsorption properties of activated carbons obtained from a pitch containing iron derivatives. *Carbon*, **39**, 2173–2179.

143. Yamashita, H., Tomita, A., Yamada, H., Kyotani, T., and Radovic, L. R. (1993). Influence of char surface chemistry on the reduction of nitric oxide with chars. *Energy Fuels*, **7**, 85–91.

144. Illan-Gomez, M. J., Linares-Solano, A., Radovic, L. R., and Salina-Martinez de Lecea, C. (1996). NO reduction by activated carbons. 7. Some mechanistic aspects of uncatalyzed and catalyzed reaction. *Energy Fuels*, **10**, 158–168.

145. Illan-Gomez, M. J., Linares-Solano, A., Radovic, L. R., and Salina-Martinez de Lecea, C. (1995). NO reduction by activated carbons. 2. Catalytic effect of potassium. *Energy Fuels*, **9**, 97–103.

146. Levasseur, B., Gonzalez-Lopez, E., Rossin, J. A., and Bandosz, T. J. (2011). Effect of reduction treatment on copper modified activated carbons on NO_x adsorption at room temperature. *Langmuir*, **27**, 5354–5365.

147. Kaneko, K. (1987). Anomalous micropore filling of nitric oxide on a-iron hydroxide oxide-dispersed activated carbon fibers. *Langmuir*, **3**, 357–363

148. AlcaÇiz-Monge, J., Bueno-Lopez, A., Lillo-Rodenas, J. M., Illian-Gomez, M. J. (2008). NO adsorption on activated carbon fibers from iron-containing pitch. *Micro. Meso. Mater.*, **108**, 294–302.

149. Bashkova, S., and Bandosz, T. J. (2011). Adsorption of NO_2 at room temperature on iron-containing polymer-based porous carbons. *Chem. Sus. Chem.*, **4**, 404–412.

150. Pietrzk, R., and Bandosz, T. J. (2007). Activated carbons modified with sewage sludge derived phase and their application in the process of NO_2 removal. *Carbon*, **45**, 2537–2546.

151. Kante, K., Deliyanni, E., and Bandosz, T. J. (2009). Adsorption of NO_2 on activated carbons modified with cerium, lanthanum and sodium chlorides. *J. Hazard. Mater.*, **165**, 357–365.

152. Bashova, S., Seredych. M., Pietrzak, R., and Bandosz, T. J. (2010). Interactions of NO_2 and NO with carbonaceous adsorbents containing silver nanoparticles. *Langmuir*, **26**, 9457–9464.

153. Alves, B. R., and Clark, A. J. (1986). An examination of the products formed on reaction of hydrogen cyanide and cyanogens with copper, chromium (6+) and copper-chromium (6+) impregnated activated carbons. *Carbon*, **24**, 287–294.

154. Alder, J. F., Fielden, P. R., and Smith, S. J. (1988). The adsorption of hydrogen cyanide by impregnated activated carbon cloth. Part I: Studies on cobalt and nickel acetates as impregnants for hydrogen cyanide removal. *Carbon*, **25**, 701–711.

155. Alder, J. F., Fielden, P. R., and Smith, S. J. (1988). The adsorption of hydrogen cyanide by impregnated activated carbon cloth. Part II. Reactivity of impregnated metal carboxylates towards hydrogen cyanide. *Carbon*, **26**, 713–721.

156. Rajakovic, L. V., Ilic, M. R., Jovanic, P. B., and Radosevic, P. B. (1995). Stoichiometric analysis of chemisorption of hydrogen-cyanide onto activated carbon cloth. *Carbon*, **33**, 1433–1441.

157. Seredych, M., Van der Merve, M., and Bandosz, T. J. (2009). Effects of surface chemistry on the reactive adsorption of hydrogen cyanide on activated carbons. *Carbon*, **47**, 2456–2465.

158. Quinn, R., Dahl, T. A., Diamond, B. W., Toseland, B. A. (2006). Removal of arsine from synthesis gas using a copper on carbon adsorbent. *Ind. Eng. Chem. Res.*, **45**, 6272–62788.

159. Hickey, J. W., and Wiig, E. O. (1948). A study of the distribution of arsine in impregnated charcoal by means of radioactive tracers. *J. Am. Chem. Soc.*, **70**, 1574–1579.

160. Petit, C., Peterson, G. W., Mahle, J., and Bandosz, T. J. (2010). The effects of oxidation on the surface chemistry of sulfur-containing carbons and their arsine adsorption capacity. *Carbon*, **48**, 1779–1787.

161. Seredych, M., Peterson, G. W., Mahle, J., and Bandosz, T. J. (2010). Interactions of arsine with nanoporous carbons role of heteroatoms in the oxidation process at ambient conditions. *J. Phys. Chem. C*, **114**, 6527–6533.

162. Katoh, H., Kuniyoshi, I., Hirai, M., and Shoda, M. (1995). Studies of the oxidation mechanism of sulphur-containing gases on wet activated carbon fibre. *Appl. Catal. B: Environ.*, **6**, 255–262.

163. Dalai, A. K., Tollefson, E. L., Yang, A., and Sasaoka, E. (1997). Oxidation of methyl mercaptan over an activated carbon in a fixed-bed reactor. *Ind. Eng. Chem. Res.*, **36**, 4726–4732.

164. Turk, A., Sakalis, E., Lessuck, J., Karamitsos, H., and Rago, O. (1989). Ammonia injection enhances capacity of activated carbon for hydrogen sulfide and methyl mercaptan. *Environ. Sci. Technol.*, **23**, 1242–1249.

165. Shin, C. S., Kim, K. H., Yu, S. H., and Ryu, S. K. (2001). Adsorption of methyl mercaptan and hydrogen sulfide on impregnated activated carbons. Presented at *7th International Conference on Fundamental of Adsorption*, Nagasaki, Japan, May 20–25.

166. Tanada, S., Boki, K., and Matsumoto, K. (1978). Adsorption properties of methyl sulfide and methyl disulfide on activated carbon, zeolite, and silicate, and their porous structure. *Chem. Pharm. Bull.*, **26**, 1527–1532.

167. Tsutsui, S., and Tanada, S. (1987). Adsorption of hydrogen sulfide, dimethyl sulfide, and their binary mixture into pores of N-containing activated carbon. *Chem. Pharm. Bull.*, **35**, 1238–1242.

168. Bashkova, S., Bagreev, A., and Bandosz. T. J. (2002). Adsorption of methyl mercaptan on activated carbons. *Environ. Sci. Technol.*, **36**, 2777–2782.

169. Karchmer, J. H. (1970). *The Analytical Chemistry of Sulfur and Its Compounds*, Vol. 1, p. 466, Wiley, New York.

170. Bashkova, S., Bagreev, A., and Bandosz, T. J. (2005). Catalytic properties of activated carbon surface in the process of adsorption/oxidation of methyl mercaptan. *Catal. Today*, **99**, 323–328.

171. Bashkova, S., Bagreev, A., and Bandosz, T. J. (2002). Dual role of water in the process of methyl mercaptan adsorption on activated carbons. *Langmuir*, **18**, 8553–8559.

172. Petit, C., Mendoza, B., and Bandosz, T. J. (2010). Reactive adsorption of ammonia on Cu-based MOF/graphene composite. *Langmuir*, **26**, 15302–15309.

173. Petit, C., Mendoza, B., Bandosz, T. J. (2010). Hydrogen sulfide adsorption on metal-organic frameworks and metal-organic frameworks/graphite oxide composites. *Chem. Phys. Chem.*, **11**, 3678–3684.

174. Levasseur, B., Petit, C., and Bandosz, T. J. (2010). Reactive adsorption of NO_2 on copper-based MOF and Graphite oxide-MOF composite. *ACS Appl. Mater. Interfaces*, **2**, 3606–3613.

Chapter 8

Conventional and Tailored Activated Carbons for Removing Natural Organic Matter and Targeted Compounds from Drinking Water

Fred S. Cannon and Cesar Nieto-Delgado

Civil and Environmental Engineering Department,
The Pennsylvania State University, 212 Sackett Engineering Building,
University Park, PA 16802, USA

fcannon@engr.psu.edu

8.1 Introduction

8.1.1 Activated Carbon

Activated carbons can remove natural organic matter (NOM) and targeted compounds from potable water sources, and thus they offer an important process for purifying drinking water. Activated carbons host an extremely high proportion of very small pores; the pores have dimensions that are similar to those of targeted compounds. These pores create a sorbed phase that can host very high concentration pockets of targeted compounds. These sorbed concentrations can be five to six orders of magnitude higher than in the aqueous phase that surrounds the activated carbon [1].

Green Carbon Materials: Advances and Applications
Edited by Thomas E. Rufford, Denisa Hulicova-Jurcakova, John Zhu

ISBN 978-981-4411-13-4 (Hardcover), 978-981-4411-14-1 (eBook)
www.panstanford.com

With their high sorption energies and fast diffusion rates, activated carbons can quickly scavenge undesirable compounds out of the aqueous phase, and contact times of minutes suffice for this sorption transfer. Moreover, the sorption onto activated carbon is achieved with no loss of product water into a retentate brine (as in membranes). Because of these factors, activated carbons serve a very important role in water treatment.

Potable water purveyors employ activated carbons that are either powdered or granular. Powdered activated carbon (PAC) (<45 microns) is generally used when municipalities experience intermittent or infrequent episodes of water contamination, such as from a chemical spill on a river, or from seasonal algal blooms that release such odorants as geosmin or methylisoborneol. Perhaps a third to a half of the US surface water treatment plants include provisions for introducing PAC when needed. Granular activated carbon (GAC) conventionally hosts grain diameters of 0.4 to 2.4 mm and are used in packed beds. Granular activated carbon is commonly employed by purveyors that experience frequent water contamination that exceeds the primary drinking water standards and/or exceeds the customer's desired thresholds. When treating surface water, GAC can be used as post-filter adsorbers, where their role is to sorb contaminants that had not been removed by the preceding filters. Alternatively, the GAC media can serve as filter bed adsorbers where the GAC offers the dual role of sorbing molecular contaminants plus filtering colloidal or particulate contaminants. Several hundred US surface water municipalities employ GAC in filter-bed adsorbers or post filter adsorber beds.

8.1.1.1 Influence of carbon source on physical-chemical properties

Activated carbons can be made from a number of parent carbon sources, and the common commercially available sources include coconut shells, hardwoods, softwoods, anthracite coal, bituminous coal, sub-bituminous coal, and lignite coal. The carbon source origin can influence the pore volume distributions, as presented for representative activated carbons in Fig. 8.1. Activated carbons derived from coconut shells and anthracite conventionally host solely micropores (width < 20 Å), while activated carbons from lignite coal conventionally host mostly mesopores (20 Å < width

< 500 Å). Bituminous coal and hardwoods offer both micropores and mesopores. However, these pore volume distributions can also be altered by tailoring protocols, as depicted for a methane-steam tailored lignite, in Fig. 8.1.

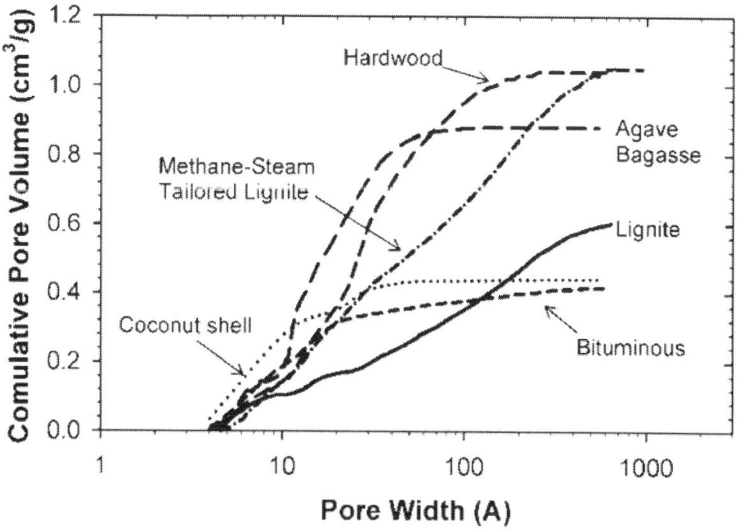

Figure 8.1 Representative pore size distribution of activated carbons derived from various source materials [2, 3].

Activated carbons comprise graphene planes stacked somewhat ordered upon one another, where each plane comprises the chicken-wire fence array of connected aromatics (see the previous sections in this chapter). The periphery of these graphene planes can include hydrogen, or surface oxides such as carboxyls, lactones, carbonyls, and phenolic hydroxyls [4]. However, not all peripheral carbons host hydrogen, oxygen, or any other substituent atom, as argued by Radovic and Bockrath [5] and Padak and Wilcox [6]. Also, the activated carbons can include nitrogen—either peripherally or within the graphene ring structure.

The activated carbon's functionality significantly impacts the sorption of contaminants: Hydrophobic contaminants will sorb preferentially onto hydrophobic surfaces, whereas negatively polar/charged species will sorb preferentially onto positively charged surfaces. In general, higher carbon content corresponds to broader graphene planes, as for the higher rank coals and coconut carbons. For example, a bituminous GAC that by proximate analysis is 90% C, 0.54% H, 0.64% N, and 8.8% O [7] hosts the empirical formula $C_{14}HON_{0.08}$. This could be mathematically computed to geometrically comprise graphene planes that on average span widths of 10–15 benzene rings (20–40 Å diameters), with most of the peripheral armchair sites occupied by carbonyls and hydroxyls (or half by carboxyls). When well-ordered, the graphene planes have interplanar spacing of 3.35 Å. However, the graphene planes may be arranged somewhat disordered relative to one another, so as to create wider pore spacing.

8.1.1.2 Thermal activation and reactivation

Activated carbons can be made from coals or lignocellulosics by a three-step conventional thermal process that includes: (a) drying the raw carbon source, (b) pyrolyzing the carbon at 800–1000°C, so as to release volatiles and graphitize the carbon, and (c) endothermically oxidizing the carbon in steam and carbon dioxide at 800–1000°C, so as to somewhat widen some of the narrow micropores into the mesopore range. For wood-based activated carbons, phosphoric acid-aided activation can be used instead, and this treatment is conducted at lower temperatures (450–650°C), so as to avoid pyrolytically collapsing the wood's pore structure.

After GAC has processed water for a while, it becomes exhausted (spent) in its capacity to sorb organic contaminants. Then the municipality may seek to thermally reactivate the carbon so that it can be used again. If conducted carefully, the thermal reactivation protocol can yield a product that removes contaminants for nearly as long as the yet-unused (virgin) carbon did. The thermal reactivation process uses similar protocols as for the initial activation. The goal of reactivation is to volatilize and gasify the sorbed organic compounds, while leaving in-tact the skeletal GAC structure. This is conceptually no simple task, in that both the skeletal GAC and the sorbed organic compounds comprise carbon, which burns. However, favorably, the sorbed organics are often more disordered and oxygenated than the skeletal GAC, and thus more susceptible to oxidative gasification. As a further confounding factor, calcium can complex with the NOM, as the Ca and NOM together sorb into the GAC. During thermal reactivation, this calcium can subsequently catalytically broaden micropores into mesopores, and the extent of such broadening should be carefully controlled [8, 9].

8.1.2 Characteristics of Contaminants That Are Targeted for Adsorption

The contaminants of interest herein that raw waters can contain include NOM, algal odorants, endocrine-disrupting compounds (EDCs), pharmaceutical and personal care products (PPCPs), volatile organic compounds (VOCs), halogenated organic compounds, and oxyanions. NOM represents the host of organic compounds that are derived from partially decomposed vegetation and microorganisms. NOM can react with chlorine disinfectants so as to form trihalomethanes (THMs), haloacetic acids (HAAs), and other disinfection byproducts (DBPs) to an extent that could exceed the primary drinking water standards. Both the NOM, odorants, and EDCs/PPCPs are often aromatics that host negatively polar functionality, and they often have moderate molecular weights. When processing groundwater, GAC is often used to remove volatile and/or halogenated organic contaminants. These include low molecular weight petrochemicals and chlorinated solvents, often of industrial and commercial origin.

8.1.2.1 Natural organic matter

The NOM commonly found in drinking water sources host molecular weights of 100–5000 Daltons, with mean MW commonly in the 600–1100 Da range [10–12]. Compounds with MW of 600–1100 have dimensions of 8 to 14 Å [8, 13]. In the presence of Fe or Al coagulant, the hydrodynamic dimensions of these can be somewhat diminished [14–16]. The NOM is often characterized as aromatics with hydroxyl, carboxyl, or methoxy ($-OCH_3$) substituents that are reminiscent of their tannin, hardwood (syringyl), and softwood (guaiacyl) plant sources [14–31]. NOM may also include long-chain alkyl carboxylated surfactants that are reminiscent of the cell membranes of algae and bacteria [30]. Several authors [17–21, 27–29] have characterized NOM by size exclusion separation through membranes and onto variously charged resins. In appraisal of five source waters, these authors [17, 27–29] have observed that 31–59% of NOM could be characterized as aquatic humic substances that were hydrophobic acids, and 5–40% were hydrophilic acids (Table 8.1). Much of the balance was hydrophobic neutrals (0–12%), hydrophilic neutrals (0.3–8%), and nitrogen-containing bases (0–5%).

Table 8.1 NOM fractions and representative structures for five source waters as discerned by resins adsorbent separation [17, 27–29]

Hydrophobic acids (31–59%)

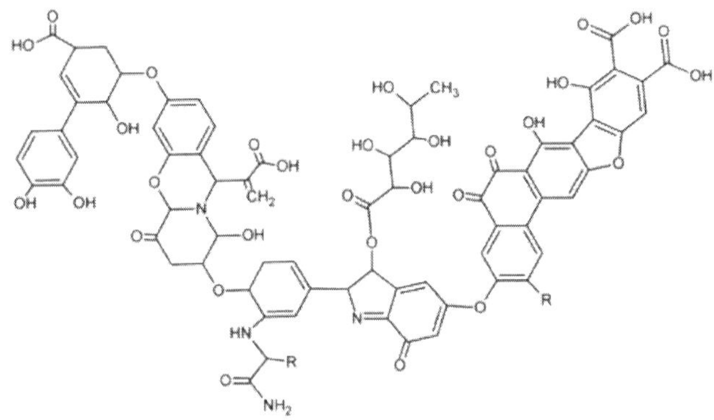

Aquatic humic substances

Hydrophilic acids (31–59%)

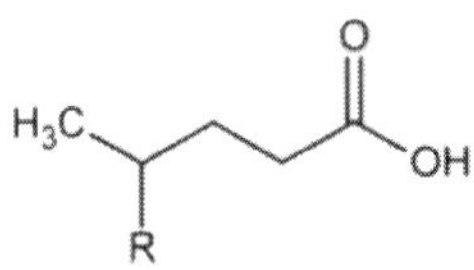

Fatty acids

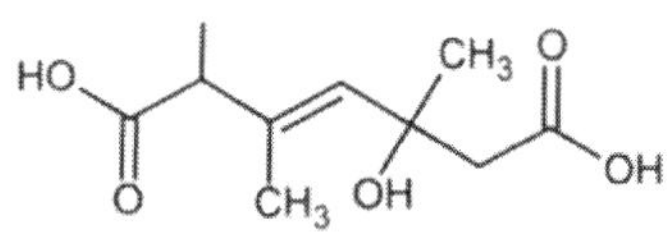

Hydroxyl acids

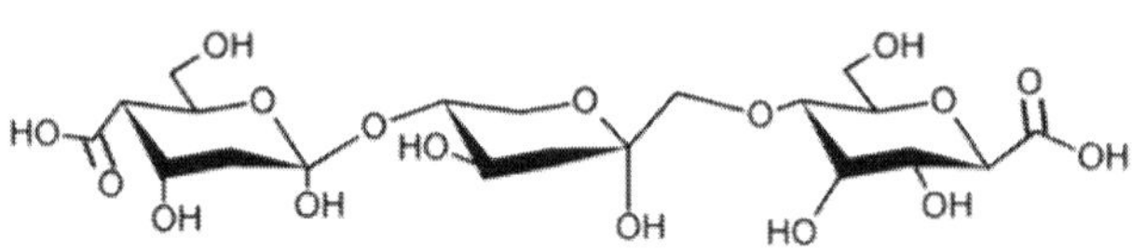

Sugar acids

Hydrophobic neutrals (0–12%)

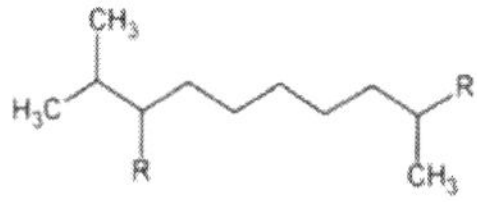

Hydrocarbons

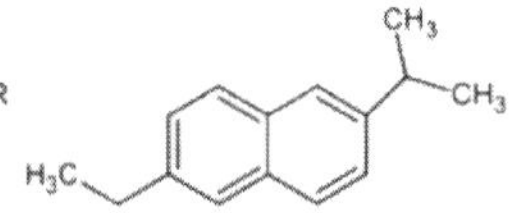

Pesticides

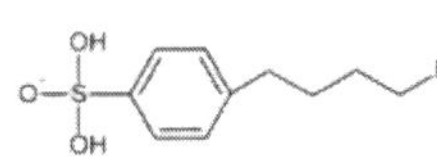

Lineal alkylbenzene sulfonates

Hydrophilic neutrals (0.2–8%)

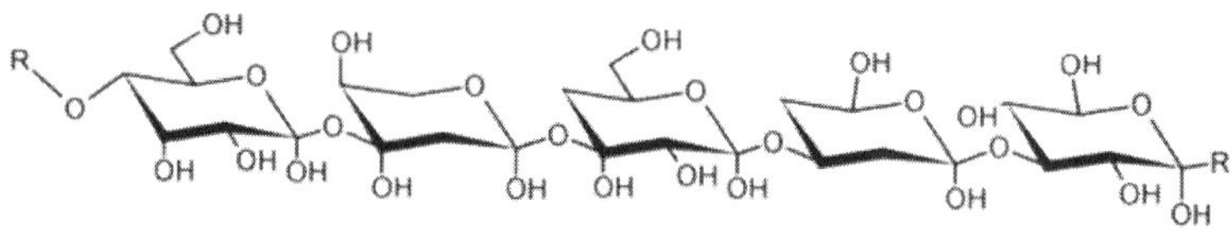

Polysaccharides

Nitrogen-containing bases (0–5%)

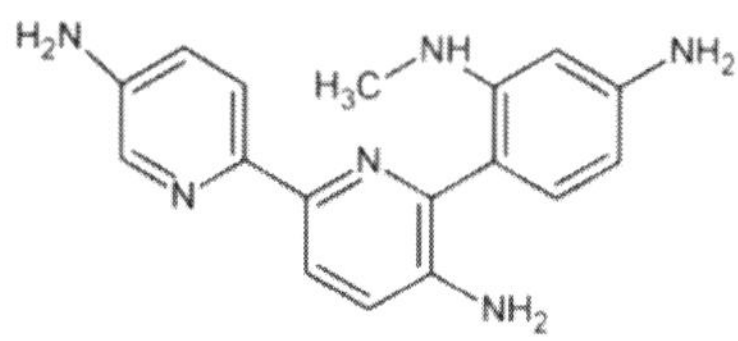

Aromatic amines

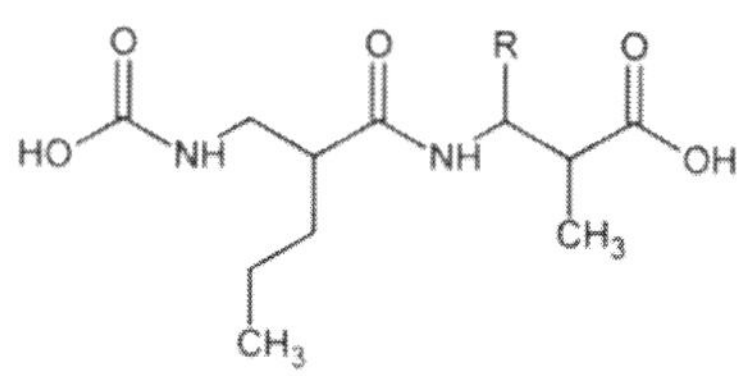

Proteins

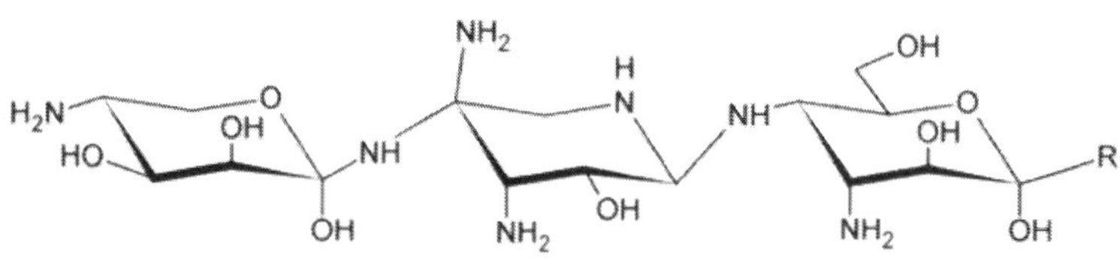

Amino sugars

Woody unit structures

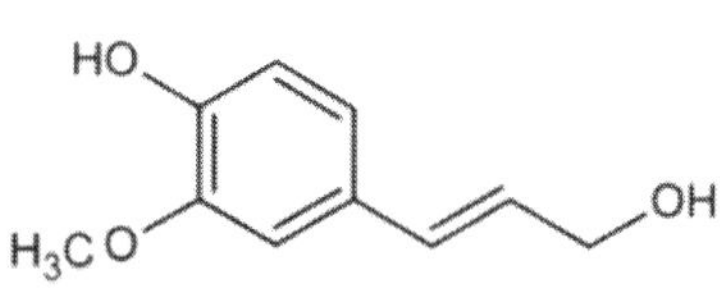

Syringyl (softwood)

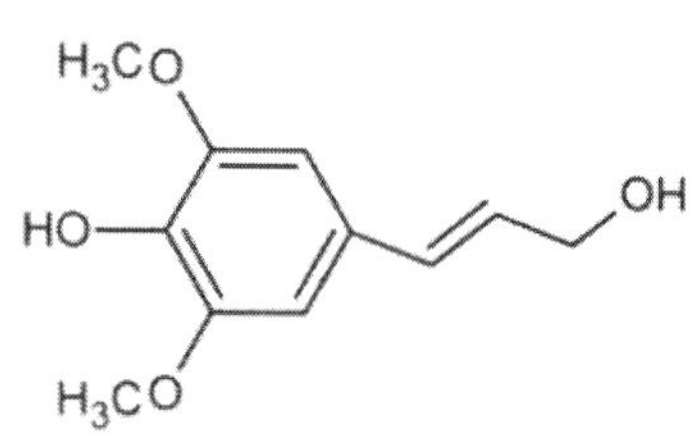

Guaiacyl (hardwood)

Much of this NOM is somewhat adsorbable onto conventional activated carbons, particularly the neutrals and bases. However, hydrophilic acids and hydrophobic acids host negative charge/polarity at neutral water pH, and these incur somewhat of a sorption limitation. This is particularly the case when the activated carbon also hosts a slight negative charge. However, to overcome this, the activated carbons can be tailored to host a net positive charge, so as to better remove this negatively charged NOM, as discussed below.

Common concentrations of NOM range from 1 to 8 mg/L as total organic carbon (TOC) in surface waters [32] and from 0.3 to 2.0 mg/L as TOC in groundwaters. The TOC is a measure of the carbon present in these NOM compounds. The carbon represents about half to two-thirds of the NOM's total molecular mass, with the balance being mostly oxygen, hydrogen, and some nitrogen (refer to NOM chemical structures, Table 8.1). The NOM itself is not generally construed as a health risk at these concentrations. However, NOM is ubiquitous in all natural waters, and whenever GAC or PAC treatment is practiced, NOM sorption will occur; the NOM sorption will compete with the sorption of other targeted compounds. Moreover, NOM can react with the disinfectant chlorine to form THMs or HAAs, for which the primary drinking water standards are 80 μg/L and 60 μg/L, respectively. Higher levels of THMs and HAAs result from higher NOM concentrations, higher chlorine concentrations, higher water temperature, and longer incubation time. This incubation time often proceeds in the water distribution system. It is generally more effective to remove the NOM DBP precursors than it is to remove the THMs and HAAs after they have formed. Without mitigating measures, quite a few surface water purveyors would exceed the THM standard, and GAC treatment offers purveyors one unit operation that will effectively diminish NOM, so as to preclude the formation of THMs and HAAs. GAC treatment is particularly effective when it is coupled with other unit operations (see below).

8.1.2.2 Geosmin and methyl isoborneol: algal bloom odorants

The odorants geosmin and methyl isoborneol (MIB) originate from seasonal algal blooms, which may persist for weeks or months—often in the summer for temperate climates. MIB and geosmin incur an unpleasant musty odor that can be detected by humans at 4 to 10 ppt. Although these odorants incur no known human health risks, their presence undermines the consumer's confidence in the water supply. For many surface water treatment plants, it is the unpredictable intermittent occurrence of MIB and geosmin that causes purveyors to include a PAC feed system in their plant design. MIB and geosmin have molecular weights of 168 and 182 Da (Tables 8.2 and 8.3), and dimensions of 5.4 to 8 Å. These compounds are quite readily adsorbable onto properly tailored activated carbons (see below), although NOM can exhibit sorption-competition [33, 34].

8.1.2.3 Endocrine-disrupting compounds, pharmaceutical, and personal care products

With emerging highly sensitive analytical instruments, we have come to realize that the a host of EDCs and PPCPs that humans use can become released into wastewater effluents, and then reenter into downstream potable water sources [35–41]. Endocrine-disrupting compounds are defined as exogenous agents that interfere with the synthesis, secretion, transport, binding, action, or elimination of natural hormones in the body that are responsible for the maintenance of homeostasis, reproduction, development, and/or behavior [39].

Hundreds of EDCs/PPCPs have been observed at ppt levels in many of the large municipal water sources that have been tested, and their human health effects at these ppt levels remain unknown. Many of these EDCs/PPCPs are 1–3 ring aromatics that have moderate molecular weights of 200–600 Daltons, with negative charge/polarity from oxygenated functionality (Tables 8.2 and 8.3). Some of these also include nitrogen in substitutents and ring structures, which can impart a positive polarity and charge.

Several of the more common and representative EDCs/PPCPs appear in Tables 8.2 and 8.3. Within the 20 source waters analyzed by Snyder [42], the most commonly observed EDCs/PPCPs included dilantin, meprobamate, sulfamethoxazole, atrazine, and carbamazepine (see Tables 8.2 and 8.3). The median concentrations of these have been in the 10–140 ppt range. Within 139 source waters analyzed by Kolpin et al. [35] and Focazio et al. [41], the most common EDCs/PPCPs were cropostanol (in 86% of waters, fecal steroid), cholesterol (84%, biogenic steroid), *N,N*-diethyltoluamide (74%, insect repellant DEET), caffeine (71, stimulant) and triclosan (58%, antimicrobial).

Table 8.2 Chemical structures of odorants and EDCs/PPCPs of concern

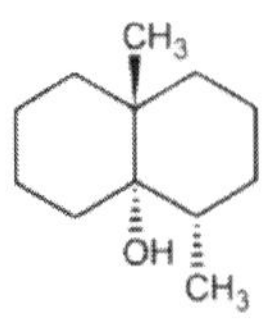	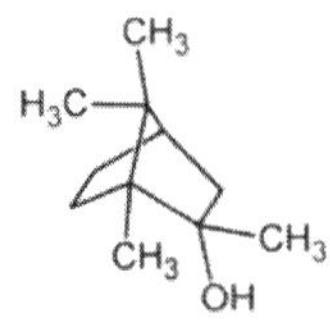	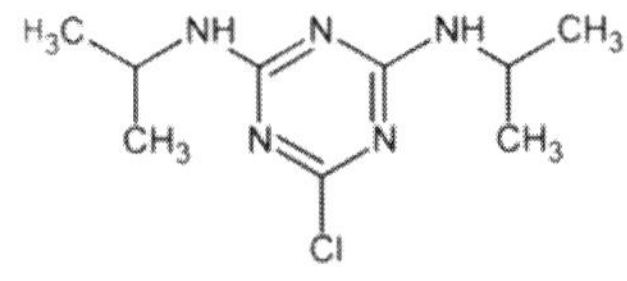
Geosmin	2-Methylisobornel	Antrazine
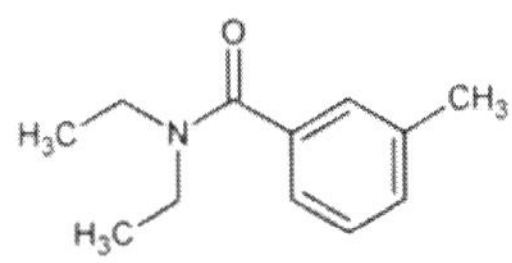	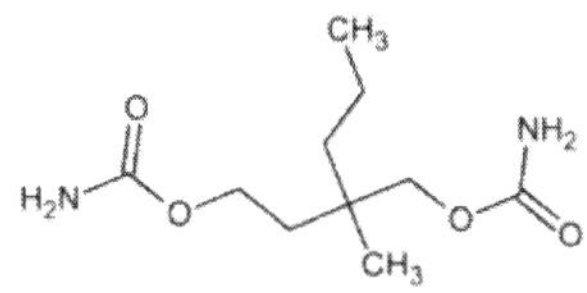	
N,N-Diethyl-meta-Toluamide (DEET)	Meprobamate	
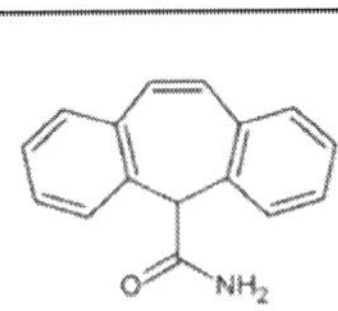	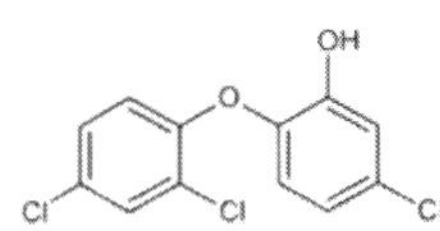	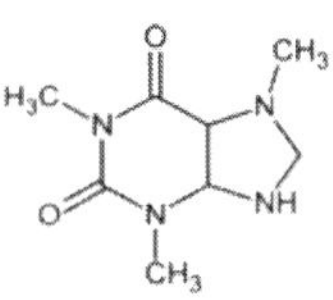
Carbamazepine	Triclosan	Caffeine
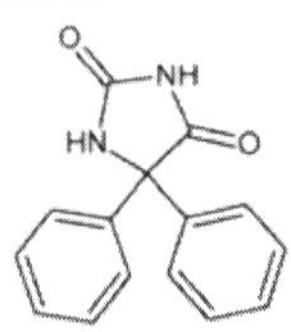	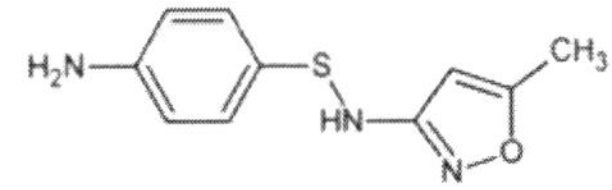	
Dilatin	Sulfamethoxazole	
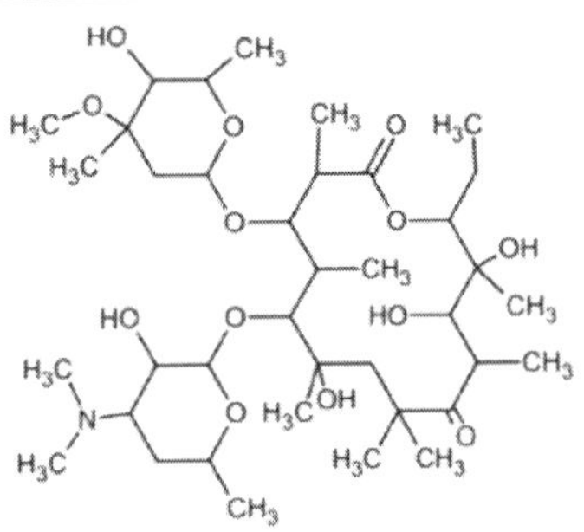	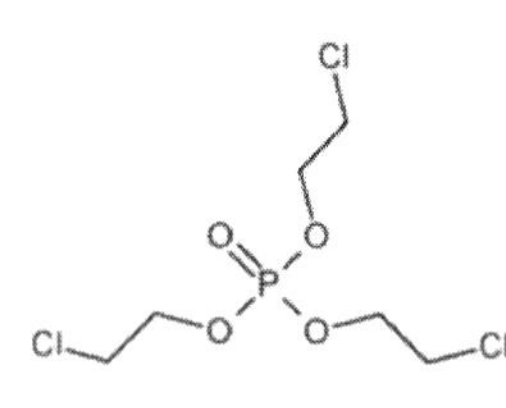	
Erythromycin	*Tris*(2-carboxyethyl) phosphine	

Table 8.3 Chemical characteristics and frequency of detection of odorants and EDCs/PPCPs*

Name	Molecular weight (Da)	$^8\chi_p$	FOSA (Å^2)	Molecular volume (cm^3/mol)	Frequency of detection	Median conc. (ng/L)**
EDC/PPCP						
Dilantin	252.3	1.299	0.0	69.47	18/20[a]	13
Meprobamate	218.3	0.167	234.4	53.73	18/20[a]	10
Sulfamethoxa-zole	253.3	0.492	96.4	64.83	18/20[a]	20
Atrazine	215.7	0.511	314.7	60.43	17/20[a]	44
Carbamazepine	236.3	1.748	43.2	69.07	16/20[a]	19
DEET	191.3	0.245	292.9	59.99	103/139[b]	60
Caffeine	194.1	0.506	239.0	49.17	98/139[b]	81
TCEP	285.5	0.188	241.0	56.13	80/139[b]	100
Triclosan	287.5	0.890	0.0	67.24	80/139[b]	140
Erythromycin	734.5	3.239	802.5	189.2	30/139[b]	100
Taste and odor						
Geosmin	182.3			184.8		4–100
MIB	168.3			173.7		4–100

Note: FOSA: hydrophobic portion of solvent-accessible surface area; $^8\chi_p$: 8th-order simple-path Chi index; DEET: *N,N*-diethyl-metatoluamide; TCEP: *tris*(2-chloroethyl) phosphate; MIB: 2-Methylisoborneol.

*As computed by Redding et al. [43]

**Median concentration when present.

[a]Snyder et al. [42].

[b]Koplin et al. [35].

When surveying the NOM, odorants, and EDCs/PPCP structures in Tables 8.1–8.3, we note that many of these compounds comprise aromatics with oxygenated substituents that pose a negative charge/polarity. For such compounds, activated carbons that offer the greatest removal capacity would be tailored to host a net positive polarity/charge (see discussion below).

8.1.2.4 Groundwater contaminants

Groundwaters can be contaminated by volatile petrochemicals, such as benzene, toluene, and xylene (BTX)—with their molecular weights of 90–120 Da and dimensions of 5–8 Å. These BTX volatiles are generally quite adsorbable, as inferred from

Freundlich isotherm parameters [44, 45] (Table 8.4). Groundwaters can also be contaminated by chlorinated solvents, such as dichloroethane, chloroform, trichloroethane, tetrachloroethane, carbon tetrachloride, trichloroethylene, and tetrachloroethylene (listed in order of increasing sorbability). The sorbability of chlorinated solvents generally increases with longer carbon chain length, more C=C double bonding, more chlorines, and more bromines (rather than chlorines) (Table 8.4). The halogen substituents incur a negative polarity on these compounds. The activated carbon vessels used for removing BTX and chlorinated solvents from groundwater are generally pressure vessels (10–30 psi); this offers the opportunity to use smaller-grained GAC in these vessels. Perhaps thousands of groundwater purveyors employ GAC for removing volatiles and chlorinated compounds from groundwater.

Table 8.4 Freundlich K and $1/n$ parameters for representative organic compounds [44, 45]

Compound	K (mg/g)(L/mg)$^{1/n}$	$1/n$
PCP	14,100	1.03
Anthracene	376	0.7
Atrazine	289	0.29
o-Xylene	174	0.47
Tetrachloroethylene	143	0.52
Toluene	97	0.43
Bromoform	92	0.66
Benzene	50	0.53
Trichloroethylene	56	0.48
Carbon tetrachloride	23	0.59
1,1,1-Trichloroethane	13	0.53
Chloroform	9	0.67
1,1-Dichloroethane	8	0.71
MTBE	6	0.48
1,2-Dichloroethane	5	5.33

Another group of compounds that can contaminate groundwaters are oxyanions, which include nitrate, sulfate, perchlorate, chromate, vanadate, arsenate/arsenite, etc. Nitrate and sulfate can appear at mg/L concentrations in native groundwaters, while the other oxyanions can appear at μg/L levels. Recent research has developed a line of tailored GACs that are well-suited for removing targeted oxyanions [46–51].

8.1.3 Mechanisms for Diffusion and Adsorption, and the Mass Transfer Zone and Bed Life

Water-phase contaminants become adsorbed into activated carbon *via* several steps: (a) advection *via* mixing (of PAC) or water flow (through GAC media), (b) molecular diffusion and then pore diffusion from the bulk water phase through the larger pores, (c) surface diffusion along pore walls and through pores that are only slightly larger than the diffusing molecules, and (d) adsorption onto the pore wall surface. Molecules that sorb along pore walls near the grains' exterior can subsequently gradually march further inward through recessed narrow pores *via* pore diffusion and surface diffusion, thus leaving peripheral sites available for yet further sorption of subsequent molecules. Molecular diffusion proceeds in the range of $0.3–3 \times 10^{-10}$ m^2/sec, which is considerably faster than for surface diffusion ($0.3–5 \times 10^{-13}$ m^2/sec). The length scales through which molecules migrate *via* pore diffusion corresponds to the grain radius. In contrast, for surface diffusion, the length scale corresponds to the radius of a "crumb" within the grain, where the "crumb" comprises closely arranged graphene planes that host very tight pores. In this context, both pore diffusion and surface diffusion can play an important role [1]. Diffusion rates are dictated by (a) the molecular size relative to the pore width, (b) specific molecule–wall interactions, (c) pore tortuosity, (d) molecular charge, (e) surface charge, and (f) water-association (i.e., the number of water molecules that are "dragged along" by the diffusing molecule).

When appraising the value and suitability of activated carbons, it is useful to consider both its carbon's adsorption capacity and its rate of adsorption. Adsorption capacity is generally dictated by the volume of pores that have widths that are 1–100 times the molecular dimensions (i.e., micropores and smaller mesopores). For example, when targeting MIB, Rangel Mendez et al. [2] found capacity dictated by pores in the 5.4–400 Å range. Concurrently, the rate of removing contaminants from the bulk phase is dictated more by the amount of pores in the larger mesopore and macropore ranges, i.e., the transport pores. When employing PAC—with its conventionally limited water contact time—it is the sorption rate that often is the more important factor. Thus, the more mesoporous carbons could be well-suited for PAC applications. Also, smaller PAC grains will adsorb faster than the larger PAC grains—since smaller grains dictate shorter diffusion lengths.

In packed-bed GAC media, the contaminants sorb in accordance with a mass transfer zone [44]. Upstream of this zone, the contaminant is completely saturated within the GAC, and below this zone, no contaminant has become present in the bulk water. For a given unit flow rate, a narrower mass transfer zone corresponds to (a) smaller and more hydrophobic molecules, (b) more transport pores, and (c) smaller GAC grains. The BTX compounds sorb with narrow mass transfer zones, since these molecules are small and hydrophobic. In contrast, NOM incurs a long transfer zone, because the molecules are larger, and comprise a wide assortment of molecules. Indeed, some portions of NOM hardly adsorb at all and bleed through the GAC immediately, while other molecules are very adsorbable.

The mass transfer zone also dictates the shape of the breakthrough curve. The breakthrough curve depicts the bed volumes (BV) of water processed versus effluent concentration. The extent of adsorption is reflected in the BV processed before the contaminant breaks through the GAC bed. Steep breakthrough curves correspond to a mass transfer zone that is well contained within the GAC media, whereas gradual breakthrough curves correspond to long mass transfer zones. For a given flow rate (normalized as the empty bed contact time—EBCT), it is the smaller GAC grains with more transport pores that can achieve steeper breakthrough curves with less premature bleeding.

Also, the adsorption capacity can also be characterized by a Freundlich isotherm of the form: $q_e = KC_e^{1/n}$, where q_e = the sorbed contaminant mass/GAC mass, C_e = the bulk contaminant mass/water volume, K characterizes the extent of sorption, and $1/n$ is the exponent. The Freundlich K and $1/n$ values have been compiled in Table 8.4 for several common and representative solvents and VOC compounds [44, 45]. Generally, K increases with increasing molecular size, C=C double-bonding, and aromaticity. The Table 8.4 data was compiled for conventional bituminous-based activated carbons that sorbed compounds in the mg/L concentration range. A number of authors (including our team) have sought to conduct Freundlich isotherm experiments as a screening-predictor of GAC bed life. However, there does not always seem to be a direct correlation, and better isotherm performance in batch mixing does not always correlate to longer BV to breakthrough in GAC beds. The distinctions could be partly related to the concentrations used: Freundlich isotherms are often performed with mg/L concentrations, while GAC beds often remove species that appear in the ppb or ppt range.

8.1.4 Parameters for Characterizing Activated Carbons, Relative to Their Performance for Adsorbing Targeted Compounds

Activated carbon manufacturers, municipalities, and scientists use an array of measurements for characterizing activated carbons. These include iodine number, apparent density, targeted-compound isotherms, dye adsorption, pore volume distributions, BET surface area, surface charge, acid titrations, Boehm titrations, slurry pH, rapid small-scale column tests (RSSCTs), and a host of atomic-scale analyses (refer to earlier chapters for the atomic-scale analyses and Boehm titrations).

The iodine number and apparent density are the most frequently used analyses employed at a carbon manufacturing factory, because these tests can be conducted simply, quickly, and in real time during production. The iodine number (mg iodine adsorbed/GAC) somewhat correlates to micropore volume and surface area for a given raw carbon source [52]. Apparent density reflects the extent to which the carbon's interior has been gasified. Typical target apparent densities range from 0.35 g/mL (for softwood carbons) to 0.49 g/mL (for bituminous coal and coconut shell carbons). Conventional wisdom dictates that for a given raw carbon source, the most favorably activated GAC will host the highest iodine number. Either under-activation, over-activation, or non-uniform activation will yield lower iodine numbers. Bituminous-based and coconut-based carbons conventionally host more micropores and higher iodine numbers than do lignite carbons; this reflects the distinct roles that these two carbon types can best offer.

Several analytical protocols have been devised for characterizing the surface chemistry of activated carbons, and these vary in their complexity and cost. Relative to conventional activated carbons, these analyses reflect the extent to which the graphene edges host carboxyls (low pH), lactones (neutral pH), or phenolics (high pH). The simplest and fastest test that offers a first-appraisal of these net proportions is the slurry pH [53]. GACs with a slurry pH above 9.5–10 will exhibit greater sorption capacity for NOM, odorants, and many EDCs/PPCPs than will GACs with a slurry pH below 6–7 [2, 43, 53]. A more extensive net analysis can be achieved with acid titrations. For this, a GAC sample is first exposed to a base, and then increments of acid are added while pH is monitored. When such titrations are normalized [33, 34], they offer insight into the carbon's surface charge versus pH. When titrations are conducted on NOM-loaded GACs, they can elucidate the acid-base character of the adsorbates.

Pore volume distributions and surface areas can be measured by placing activated carbon grains in a tube, immersing the tube in a bath of liquid argon or nitrogen, evacuating the gases within the tube, then dosing incrementally more gaseous nitrogen or argon into the tube, while pressures progress from 10^{-6} atm to nearly 1 atm over the course of 1–2 days. When immersed in this bath, the gaseous argon or nitrogen will condense within the GAC pores—with extent of condensation dictated by the pressure and width of the pores. Then, complex mathematics is employed to compute the pore volume distributions [54, 55] and BET surface area. Pore volume distributions of an array of activated carbons have been presented above in Fig. 8.1. When properly interpreted, these pore volume distributions offer a very valuable predictive tool for anticipating contaminant adsorption capacities and kinetics.

8.1.4.1 Rapid small-scale column tests

The full-scale bed life for removing a targeted contaminant can be directly simulated by RSSCTs. The RSSCT approach employs GAC grains that are crushed and sieved to a considerably smaller size than are used in full-scale GAC beds; thus this abbreviates the diffusion length scales. This can mean that month-long full-scale performance can be simulated by day-long bench-scale performance. Per Crittenden et al. [56], the RSSCT similitude can be expressed as follows:

$$\frac{\text{EBCT}-\text{LS}}{\text{EBCT}-\text{SS}}=\left(\frac{d-\text{LS}}{d-\text{SS}}\right)^{n},$$

where EBCT represents the contact time that the water would have within the GAC bed if the bed were empty, d = grain diameter, LS = large scale, and SS = small scale. The exponent n is 1 for proportional diffusivity, and 2 for constant diffusivity. Large compounds and charged molecules tend to be best modeled with n = 1; small hydrophobic compounds by n = 2. Empirically, some in-between compounds are best simulated by n = 1.5.

8.2 GAC in Water Treatment Systems in Combination with Coagulation, Advanced Oxidation and Biological GAC

In surface water treatment plants, GAC is used in combination with other unit operations, and these together achieve the primary drinking water standards regarding organic compounds and DBPs. The GAC can be used in post-filter adsorbers, as at the Cincinnati Richard Miller plant. Alternatively, the GAC can be used in filter-bed adsorbers, and we pose the Delaware River plant (NJ) as an example of that. GAC treatment performance can be enhanced when it is synergistically coupled with other water treatment unit operations.

8.2.1 Cincinnati Water Treatment Plant

Cincinnati's Richard Miller Water Treatment Plant hosts 12 beds of GAC, each of which holds 600 m^3 of GAC. Each provides a 15–20 min EBCT. The GAC beds process water that has been coagulated and filtered. The GAC is reactivated in multiple hearths about every 7–10 months. Our team [57–61] monitored the performance of three full-scale GAC beds that operated in parallel: (a) virgin, (b) reactivated 5 times, and (c) reactivated 12 times. During this study, these beds operated with an average EBCT of 20 min; so 225 days corresponded to 16,000 BV. These three GAC beds removed TOC about equally (Fig. 8.2). The TOC trends in these figures also depict commonly observed trends in TOC breakthrough: even at the beginning of the run, roughly 10–15% of the TOC was already breaking through the

GAC beds; this represented the non-adsorbable NOM. Then 225 days later, the GACs continued to remove 25% of the TOC, which represented the highly adsorbable and/or colloidal (i.e., filterable) fractions. By removing this NOM, all three of these beds also removed the potential to form trihalomethanes to below 40 μg/L, and the potential to form HAAs to below 30 μg/L throughout this trial [60]. However, bench-scale isotherms revealed that the 5× and 12× reactivated GACs removed benzene less efficiently than did the virgin GAC. Benzene is a surrogate for the types of volatiles that have been intermittently spilled on the Ohio river upstream of Cincinnati.

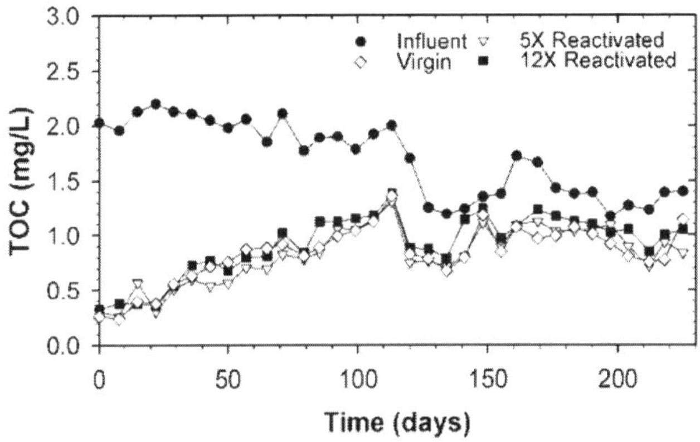

Figure 8.2 Cincinnati, OH full-scale GAC performance: TOC removal by GAC that was virgin, 5 times reactivated, and 11 times reactivated; 225 days = 16,000 BV [60].

The Penn State team conducted acid titrations of GAC samples that were taken from the three Cincinnati beds as water treatment service progressed. Figure 8.3 depicts analyses taken from the GAC beds' top 6 inches at the beginning, second month, and end of these runs. In this context, the titrations offered insight into the acidity and net negative charge of the sorbed NOM. Specifically the GACs' point of zero charge began at pH 8.7 for the virgin GAC and 9.3–9.4 for the 5× and 12× reactivated GACs. Then after 57 days of sorbing NOM, the net point of zero charge for these GACs had progressed to pH 6.3–7; then after 225 days to pH 5–5.5. Notably, the equivalents of negative charge represented by the sorbed NOM roughly balanced the equivalents of Fe, Al, and Ca build-up in these GACs [58], and this highlighted how NOM complexed with metals, and then the complexes sorbed into the GAC.

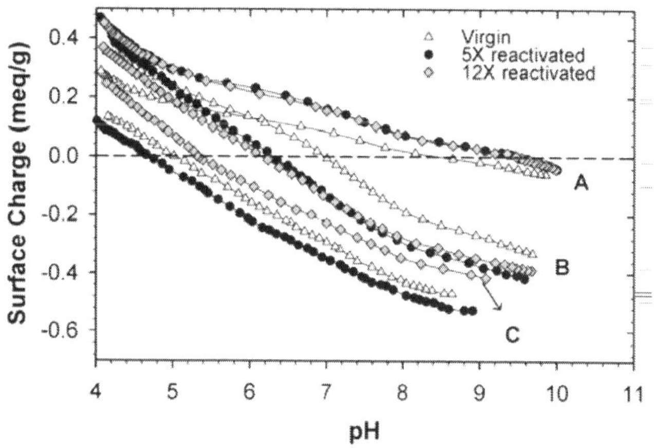

Figure 8.3 Cincinnati OH full-scale GAC; time-wise changes in net surface charge: A (top three), day 0, B (middle three), day 57, C (bottom three), day 226 [58].

8.2.2 Delaware River Regional Water Treatment Plant, New Jersey

The Delaware River Regional Water Plant, owned by the New Jersey-American system, hosts GAC filter bed adsorbers that are preceded by coagulation, sedimentation, and ozonation. The ozonation oxygenated the NOM, rendering it less recalcitrant to biological decomposition; thus ozonation has encouraged biological growth in the GAC filters. Water purveyors have sometimes (but not always) observed that with biological enhancement of GAC, the TOC removals will remain greater than without this biological enhancement. At the Delaware River plant, even after 33 months of operation without change-out, these biological GAC filters continued to remove 10–15% of the TOC.

Our research team employed tetramethylammonium thermochemolysis (TMAH) gas chromatography-mass spectrophotometry (GC-MS) to identify and quantify 30–40 specific organic compounds that appeared on these chromatograms, and these species represented 1–2.7% of the organic carbon mass present in these water samples. These species were monitored through each of the plant's unit operations [30, 62]. The chromatographically identified species included fatty acids (C8–C26), and single-ring aromatics with substituents of hydroxyl, carboxyl, methoxy, and carbonyl. This TMAH protocol added methyl groups to hydroxyls or carboxyls. With regard to the plant influent, this confounding methylation issue meant that this protocol could not distinguish whether the native NOM hosted hydroxyls or methoxy substituents (i.e., reminiscent of syringyl or guaiaquil lignin). Ozonation created considerably more short-chain fatty acids and hydroxylated aromatics (ozonation would not have created methoxy aromatics). The upflow clarifiers removed half of hydroxylated aromatics and short-chain fatty acids. The hydroxylated and carboxylated aromatics were initially removed by relatively fresh GAC (through 12,000 BV), but broke through with prolonged biological GAC operation (23,000 BV and beyond). The fatty acids had not broken through at 60,000–90,000 BV, but exhibited partial breakthrough (and/or formation-bleeding) at 112,000 BV (Fig. 8.4).

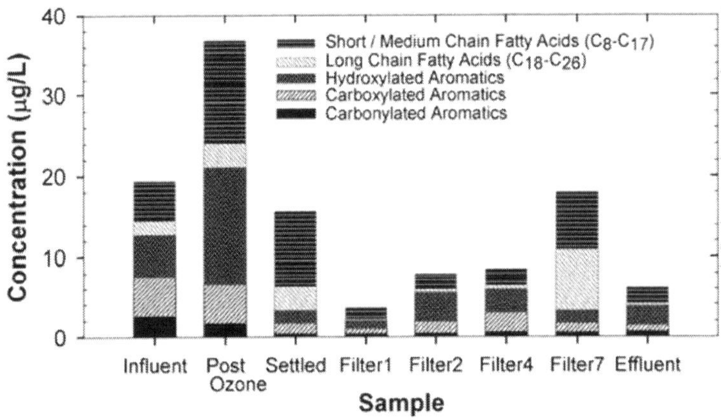

Figure 8.4 Delaware River Water Plant: Species quantified by TMAH-GC-MS through unit operations, representing 1% to 2.7% of organic mass. GAC biofilters had processed 12,000 BV (Filter 1), 23,000 BV (Filter 2), 60,000–90,000 BV (Filter 4) and 112,000 BV (Filter 7). In influent, hydroxylated aromatics were not distinguishable from methoxy aromatics [30, 62].

8.2.3 Metal Coagulants, Complexation with Natural Organic Matter

Proper metal salt coagulation and sedimentation can greatly enhance TOC removal through GAC [63, 64], as depicted by Fig. 8.5. In bench-scale tests, coagulation, clarification, and filtration diminished the influent TOC load onto the GAC media, which prolonged the bed life for removing TOC. Yet more subtly, the ferric cation also complexed with the anionic NOM to create more tightly condensed and less negatively polar compounds, which sorbed more extensively. Thus, proper ferric salt pretreatment and pH adjustment could increase GAC's TOC sorption capacity by 30–40% [63].

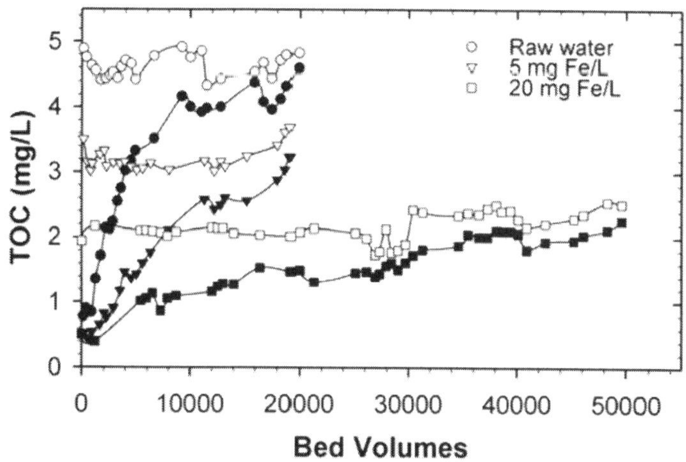

Figure 8.5 Effect of ferric chloride pretreatment on effluent TOC levels through RSSCTs (pH = 6). Influent: open symbols; effluent: filled symbols [49, 64, 65].

8.2.4 Retrofitting Conventional Gravity Filters with GAC

With progressively more constraining water quality standards, numerous municipalities have been retrofitting their conventional gravity filters with GAC filter bed adsorbers. The conventional filters often contain anthracite (1 to 3 mm grain size range), silica sand (0.4 to 0.8 mm), and sometimes ilmenite or garnet (0.1 to 0.3 mm range). The gravity filter media are commonly 2.5 to 3 feet deep; these grain sizes are conventionally selected to allow up to 6–8 feet of headloss through the media after 2–4 days of turbidity build-up on the filters between backwashes. When retrofitting such filter media, the common practice has been to replace the sand, anthracite, and ilmenite with 2.5 to 3 feet of GAC that is US Mesh #12 × 40 or #8 × 30 (i.e., in the 0.4 to 2.4 mm range). At the conventional 4 gpm/ft^2 loading, this translates to a 5 min EBCT. However, for these grain sizes, this EBCT is too short to contain the NOM mass transfer zone, and too many organic compounds break through such a bed.

The recommendation of these authors is to instead replace the conventional filter media with #20 × 50 mesh (0.3 to 0.85 mm) GAC, which has about the same dimensions as the silica sand. Also, these authors recommend that the media be made as deep as the plant's hydraulic profile will allow (for example, by also replacing underdrain gravel with porous plates). We note that per the equation governing RSSCTs posed above, a 5 min EBCT with #20 × 50 grains will contain the mass transfer zone as nicely as when using a 10 min EBCT with #12 × 40 grains that are twice as big. To yet further enhance such a retrofit, the authors also recommend that the municipalities update the coagulation and sedimentation operations so that the flocculant load and headloss build-up onto the filter bed adsorbers is minimized. Such enhancements could include installing preozonation, up-flow clarifiers, and/or prolonged sedimentation times.

8.3 Tailoring Activated Carbons for Enhanced Removal of Natural Organic Matter and Odorants

Building on the fundamental concepts conveyed above, our Penn State team has sought to tailor activated carbons that could better remove NOM, odorants, EDCs/PPCPs, halogenated organic compounds, and VOCs. For example, the Penn State team has thermally tailored a conventional lignite GAC in steam or methane/steam (with cool-down in steam) in a manner that yielded twice the pore volume distribution (see Fig. 8.1), while exhibiting a slurry pH above 9.5–10.0 [2, 53]. In RSSCTs, the more favorable tailored activated carbon offered five times the bed life for 14C-MIB (Fig. 8.6a), and 1.5 to 2 times the bed life for two THMs: chloroform and bromoform (Fig. 8.6b). Yet further, in pilot studies, this tailored lignite GAC removed all TOC for 3000 BV (Fig. 8.7). No other known GACs have removed all NOM-TOC for that duration.

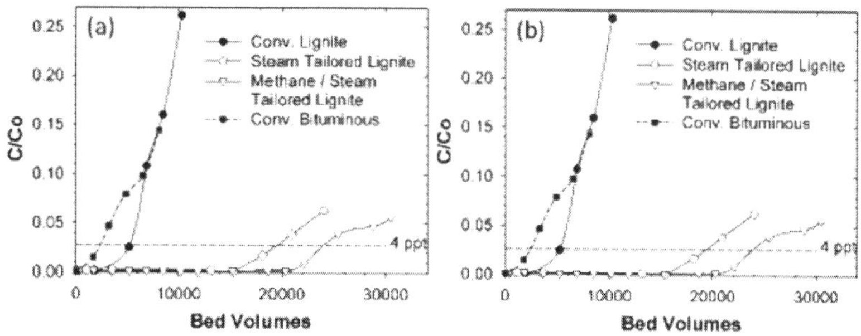

Figure 8.6 (a) RSSCTs of MIB onto conventional and tailored activated carbons [2]. (b) RSSCTs of chloroform and bromoform onto conventional and tailored activated carbons. Data provided by Hyok Bo Kwan at PSU.

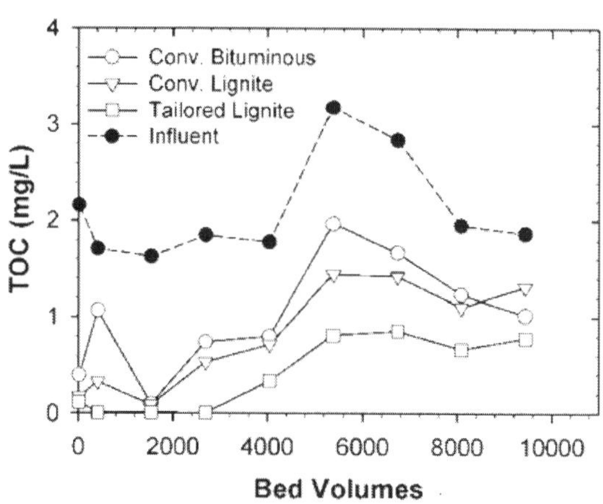

Figure 8.7 TOC removal through tailored lignite, as compared with conventional bituminous and conventional lignite GAC during pilot scale test at Passaic Valley, NJ. Data by Nowack at PSU.

A significant cause of diminished TOC capacity in conventional GACs occurs when oxygen chemisorbs onto the reactive sites along the graphene edges, causing the carbon to host a more negative charge at neutral pH—corresponding to a lower slurry pH. This oxygen can originate from water that contains dissolved oxygen during treatment service, or from ambient air during prolonged storage. For example, activated carbons stored in ambient air within a screw-tight container can drop their slurry pH from

8.5–9.0 down to 6–7 within a year. However, our team observed that methane/steam-tailored carbons or steam-tailored carbons exhibited far less propensity to chemisorb oxygen. Specifically, these thermally passivated carbons chemisorbed only 100–250 μg oxygen/g GAC when exposed to O_2 in a TGA at 105°C for 3 h, per Fig. 8.8. This compared to a nitrogen-pyrolyzed activated carbon that chemisorbed 6000 μg/g oxygen in 3 h [66]. The Penn State team perceived that this recalcitrance against oxygen chemisorption that was offered by these thermally tailored GACs represented an important factor in their prolonged bed life and complete TOC removal.

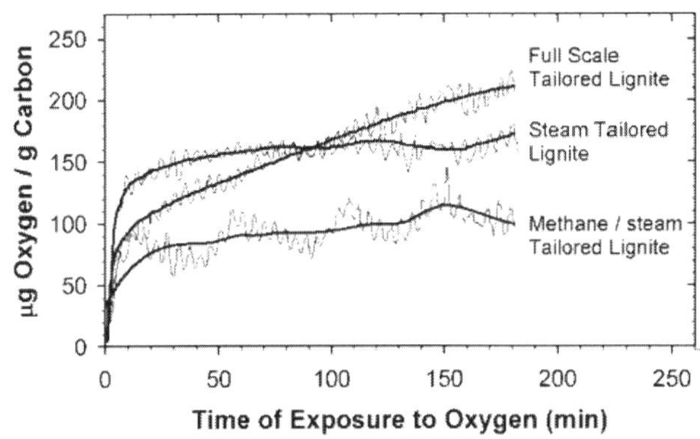

Figure 8.8 Thermo-gravimetric diagrams for oxygen uptake by tailored activated carbons. Oxygen uptake at 105°C. In comparison, an N_2-pyrolyzed GAC took up 6000 μg/g oxygen in 3 h [66]. Data by Rangel-Mendez at PSU.

8.4 Modeling Sorption of Endocrine-Disrupting Compounds and Pharmaceutical/Personal Care Products

In further trials aimed at appraising EDC/PPCP removal, our team has compared conventional lignite GAC to a steam-tailored lignite GAC, and a methane-steam tailored lignite GAC. The tailored carbons offered more mesopores, more micropores, and higher slurry pH than did the conventional lignite carbons. RSSCTs were conducted on each of these three GACs, while processing Lake Mead, NV water that had been spiked concurrently with 30 EDC/PPCPs at about 100 ppt each. For the tailored lignite carbons, BV to breakthrough ranged from 15,000 to >90,000 BV, compared with 2,000–55,000 BV for the conventional lignite carbons (Fig. 8.9). Significantly, for all EDC/PPCPs that exhibited <20,000 BV to breakthrough with the conventional lignite, the bed life increase three- to eightfold with the tailored lignites.

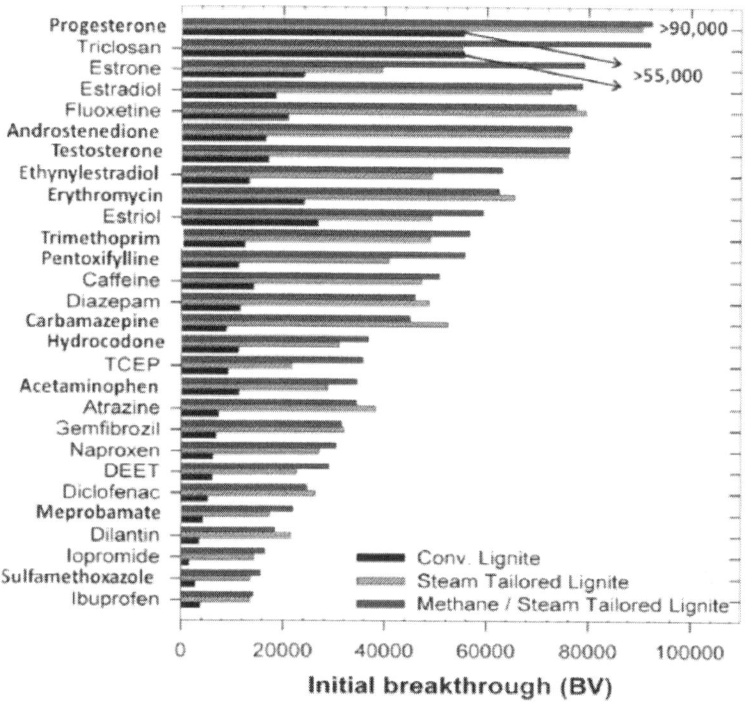

Figure 8.9 RSSCT BV processed at characteristic initial breakthrough for the 28 EDCs/PPCPs exhibiting breakthrough [43].

For these three activated carbons with these EDC/PPCP compounds, the bed life was characterized by the following expression:

$$BV_p = \left(\frac{(TPV \times \rho_{mc})(e^{0.2812 \times pH_s})}{(CV \times C0)}\right)(0.2758 \times {}^{8}\chi_p + 0.0011 \times FOSA)$$

This QSAR-like model correlated to the behavior of 22 of the 30 EDC/PPCP compounds analyzed (Fig. 8.10). The eight exceptions were compounds that did not break through or had outlier chemical characteristics. In this model, pH_S = slurry pH, and TPV = total pore volume <500 Å width. The parameters CV, FOSA and $^{8}\chi_{p}$ characterized the molecules, as per Table 8.4. These parameters were computed by Redding et al. [53] using software packages listed in [53]. Intriguingly, this bed life conformed to a volumetric-based relationship that hosted some of the features of an isotherm (Fig. 8.11). Namely, bed life could be predicted on the basis of volumetric characterizations: volume of molecules/volume of GAC pores versus volume of water processed/volume of GAC bed.

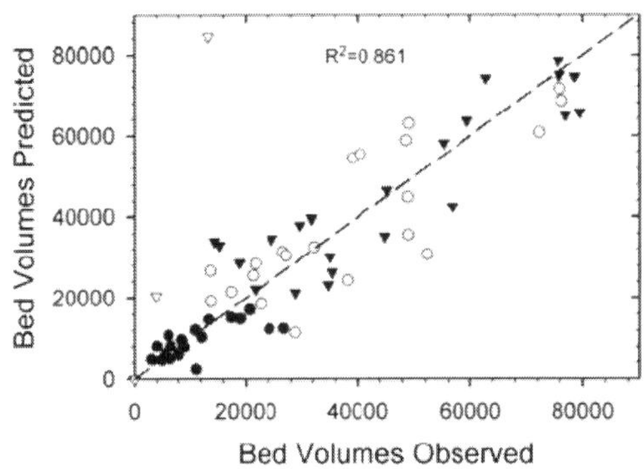

Figure 8.10 Predicted versus observed initial breakthrough for 23 EDCs/PPCPs as described by the QSAR-like model. Adsorption generally increased in this order: M/S (▼) > STEAM (○) > HD4000 (●) [43].

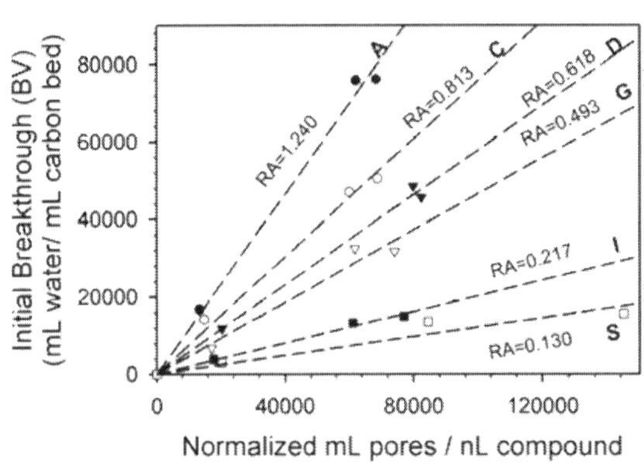

Figure 8.11 Relative Adsorbability (RA) for several compounds: A = androstenedione (•), C = caffeine (○), D = diazepam (▼), G = gemfibrozil (∇), I = ibuprofen (■), and S = sulfamethoxazole (□) [43].

8.5 GAC Tailored for Removing Oxyanions

Our Penn State team has also preloaded quaternary ammonium (QA) surfactants into GACs; the preloaded QA functionality subsequently captured perchlorate out of groundwater. In RSSCTs with groundwater that contained 30–70 ppb perchlorate, the BV to 6 ppb breakthrough reached 28,000–38,000 BV when the QA was either a trimethyl ammonium or pyridinium surfactant; the surfactant contained 12–18 carbons. Figure 8.12 summarizes these results, with more discussion of variable factors in [48, 66, 67]. These RSSCTs employed proportional diffusivity similitude. After exhaustion, roughly 10% of all QA sites were occupied by perchlorate [50]. In full-scale demonstrations, perchlorate removal bed life was about half this long, and the shortened bed life could have in part been due to competition with thiosulfate, which was representative of the somewhat-reduced sulfur species that can be found in groundwater while it remains anoxic [65, 67].

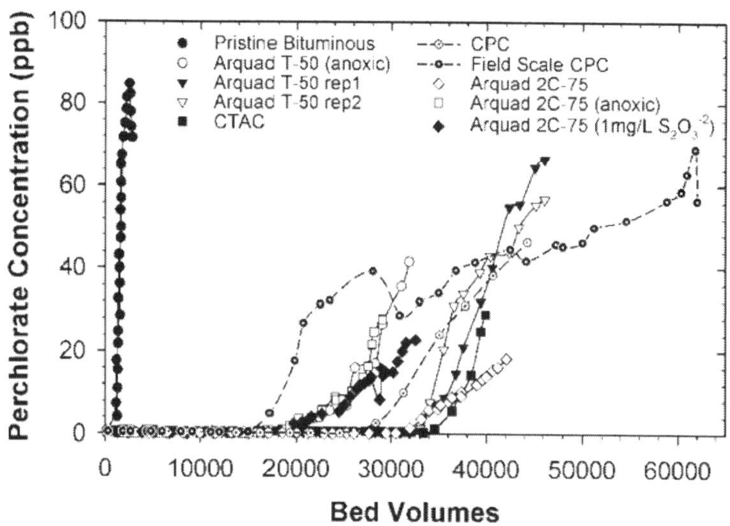

Figure 8.12 Perchlorate removal from Redlands water onto AC tailored with several cationic surfactants: for RSSCTs, except as listed as Field Scale. With well-oxygenated groundwater, except as listed with anoxic groundwater [51, 65, 67].

Our Penn State team has also preloaded GAC with iron (hydroxide) nanoparticles so as to enhance the bed life for arsenate and arsenite removal. The more favorable removals were achieved with 10–12% loading as Fe, and these iron-tailored GACs could remove arsenate/arsenite from 50 ppb influent down to <10 ppb effluent for 22,000–30,000 BV [48, 68, 69] (Fig. 8.13). With pristine GAC, As breakthrough occurred within 100 BV. For this oxyanion, the RSSCTs based on proportional diffusivity quite closely matched the pilot-scale performance (Fig. 8.13).

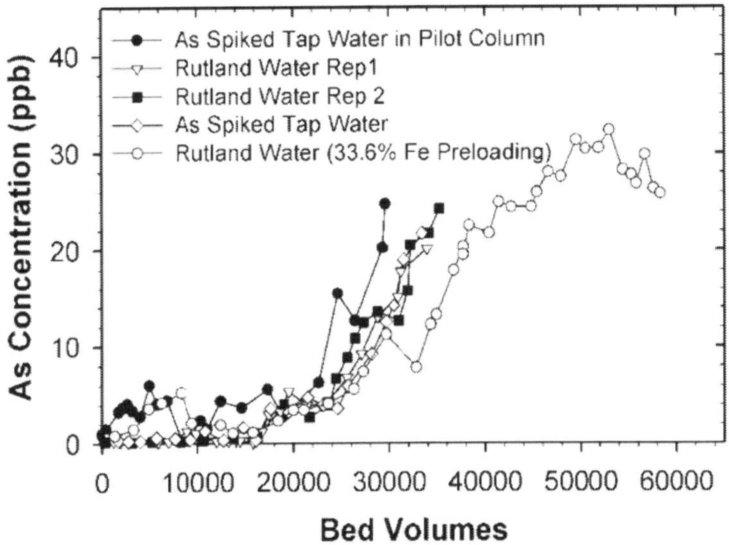

Figure 8.13 Arsenate removal by iron tailored GAC: with 11% to 12.5% Fe preloading, except as noted; for RSSCTs except as noted for pilot column [47, 48, 68, 69].

These iron nanoparticles mostly occupied pores in the 5–100 Å range (Fig. 8.14), and the nanoparticle size was somewhat dictated by the preloading protocol. Nieto-Delgado et al. [49] developed agave bagasse-based activated carbons with iron oxy-hydroxide nanoparticles, and these nanoparticles were also primarily in the 20 to 100 Å pores. The optimal iron-tailored agave GACs adsorbed 4.6 mg As/g GAC when the isotherm water contained 1.5 mg/L As initially.

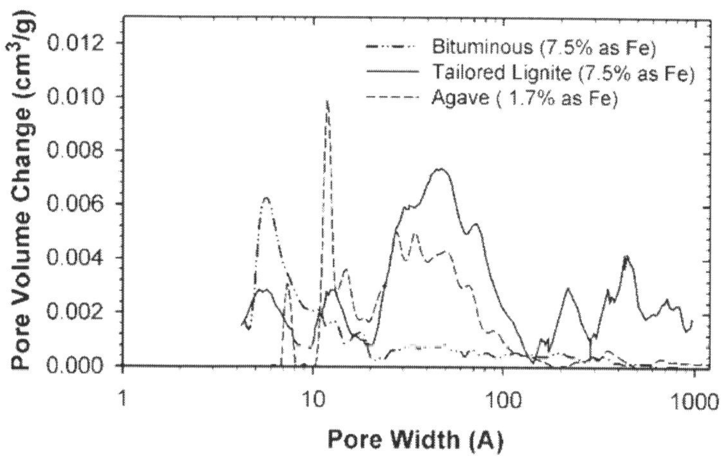

Figure 8.14 Change in incremental pore volume due to iron preloading (listed as % preloading as Fe) [48, 49, 70].

8.6 Overview

In overview, activated carbons offer a valuable process for removing contaminants that appear in natural waters. Many of the organic contaminants have aromatic structures, with oxygen-containing substituents that create negative charge/polarity. Moreover, activated carbons can be tailored in a manner that profoundly improves their bed life for removing NOM, EDCs/PPCPs, odorants, and oxyanions. For many common contaminants, the favorable tailoring is achieved by increasing the activated carbon's pore volume and positive surface charge.

References

1. Cannon, F. S., and Roberts, P. V. Activated carbon: Adsorption of DOC from wastewater. *J. Environ. Eng. Division*, 1982, **EE4**, pp. 766–783.
2. Rangel-Mendez, J. R., and Cannon, F. S. Improved activated carbon by thermal treatment in methane and steam: Physicochemical influences on MIB sorption capacity. *Carbon*, 2005, **43**(3), pp. 467–479.

3. Nieto-Delgado, C., Terrones, M., and Rangel-Mendez, J. R. Development of highly microporous activated carbon from the alcoholic beverage industry organic by-products. *Bio. Bioenergy*, 2011, **35**(1), pp. 103–112.
4. Laszlo, K., Josepovits, K., and Tombacz, E. Analysis of active sites on synthetic carbon surfaces by various methods. *Anal. Sci.*, 2001, **17**, pp. i1741–i1744
5. Radovic, L. R., and Bockrath, B. On the chemical nature of graphene edges: Origin of stability and potential for magnetism in carbon materials. *J. Am. Chem. Soc.*, 2005, **127**(16), pp. 5917–5927.
6. Padak, B., and Wilcox, J. Understanding mercury binding on activated carbon. *Carbon*, 2009, **47**(12), pp. 2855–2864.
7. Chen, W., Cannon, F. S., and Rangel-Mendez, J. R. Ammonia-tailoring of GAC to enhance perchlorate removal. I: Characterization of NH_3 thermally tailored GACs. *Carbon*, 2005, **43**(3), pp. 573–580.
8. Cannon, F. S., et al. *Effect of Calcium in Field-Spent GACs on Pore Development during Regeneration*. Vol. 85. 1993: American Water Works Association.
9. Cannon, F. S., et al. Reaction mechanism of calcium-catalyzed thermal regeneration of spent granular activated carbon. *Carbon*, 1994, **32**(7), pp. 1285–1301.
10. Chin, Y.-P., Aiken, G., and O'Loughlin, E. Molecular weight, polydispersity, and spectroscopic properties of aquatic humic substances. *Environ. Sci. Technol.*, 1994., **28**(11), pp. 1853–1858.
11. Peuravuori, J., and Pihlaja, K. Molecular size distribution and spectroscopic properties of aquatic humic substances. *Anal. Chim. Acta*, 1997, **337**(2), pp. 133–149.
12. Assemi, S., et al. Characterization of natural organic matter fractions separated by ultrafiltration using flow field-flow fractionation. *Water Res.*, 2004, **38**(6), pp. 1467–1476.
13. Perry, J. H., and Chilton, C. H. *Chemical Engineers' Handbook*. 5th ed. 1973, New York: McGraw-Hill. xv, 1942 p.
14. Bowen, W. R., and Mohammad, A. W. Characterization and prediction of nanofiltration membrane performance: A general assessment. *Chem. Eng. Res. Des.*, 1998, **76**(8), pp. 885–893.
15. Combe, C., et al. The effect of CA membrane properties on adsorptive fouling by humic acid. *J. Membr. Sci.*, 1999, **154**(1), pp. 73–87.
16. Shon, H. K., Vigneswaran, S., and Ngo, H. H. Effect of partial flocculation and adsorption as pretreatment to ultrafiltration. *AIChE J.*, 2006, **52**(1), pp. 207–216.

17. Leenheer, J. A. Comprehensive approach to preparative isolation and fractionation of dissolved organic carbon from natural waters and wastewaters. *Environ. Sci. Technol.*, 1981, **15**(5), pp. 578–587.

18. Leenheer, J. A., and Croua, J.-P. Peer Reviewed: Characterizing aquatic dissolved organic matter. *Environ. Sci. Technol.*, 2003, **37**(1), pp. 18A–26A.

19. Rosario-Ortiz, F. L., Snyder, S. A., and Suffet, I. H. Characterization of dissolved organic matter in drinking water sources impacted by multiple tributaries. *Water Res.*, 2007, **41**(18), pp. 4115–4128.

20. Imai, A., et al. Fractionation and characterization of dissolved organic matter in a shallow eutrophic lake, its inflowing rivers, and other organic matter sources. *Water Res.*, 2001, **35**(17), pp. 4019–4028.

21. Kanokkantapong, V., et al. Characterization of haloacetic acid precursors in source water. *J. Environ. Manag.*, 2006, **80**(3), pp. 214–221.

22. Saiz-Jimenez, C. Analytical pyrolysis of humic substances: Pitfalls, limitations, and possible solutions. *Environ. Sci. Technol.*, 1994, **28**(11), pp. 1773–1780.

23. de Leeuw, J. W., and Baas, M. The behaviour of esters in the presence of tetramethylammonium salts at elevated temperatures: flash pyrolysis or flash chemolysis? *J. Anal. Appl. Pyrolysis*, 1993, **26**(3), pp. 175–184.

24. Martín, F., et al. Pyrolysis derivatization of humic substances 2. Pyrolysis of soil humic acids in the presence of tetramethylammonium hydroxide. *J. Anal. Appl. Pyrolysis*, 1995, **31**, pp. 75–83.

25. Martin, F., et al. Pyrolysis derivatization of humic substances 1. Pyrolysis of fulvic acids in the presence of tetramethylammonium hydroxide. *J. Anal. Appl. Pyrolysis*, 1994, **28**(1), pp. 71–80.

26. Hatcher, P. G., and Clifford, D. J. Flash pyrolysis and in situ methylation of humic acids from soil. *Org. Geochem.*, 1994, **21**(10–11), pp. 1081–1092.

27. Pelekani, C., et al. Characterization of natural organic matter using high performance size exclusion chromatography. *Environ. Sci. Technol.*, 1999, **33**(16), pp. 2807–2813.

28. Korshin, G. V., Benjamin, M. M., and Sletten, R. S. Adsorption of natural organic matter (NOM) on iron oxide: Effects on NOM composition and formation of organo-halide compounds during chlorination. *Water Res.*, 1997, **31**(7), pp. 1643–1650.

29. Croue, J. P., Korshin, G. V., and Benjamin, M. *Characterization of Natural Organic Matter in Drinking Water.* 2000, United States: AWWA Research Foundation and AWWA Association.

30. Crosson, K. M., Frazier, S. W., Hatcher, P. G., Cannon, F. S., Casale, R., and Benesi, A. J. *Assessing the Effects of Drinking Water Treatment Processes on Refractory NOM Fractions using Tetramethylammonium Hydroxide (TMAH)-Thermochemolysis*. 2011, The Pennsylvania State University.

31. Frazier, S. W., et al. Characterization of organic matter from natural waters using tetramethylammonium hydroxide thermochemolysis GC-MS. *J. Anal. Appl. Pyrolysis*, 2003, **70**(1), pp. 99–128.

32. Voorhees, N. J. *Source Water Characterization: Impact on Process Design and Treatment Optimization*. 1993: American Water Works Service Company, Inc.

33. Newcombe, G., Drikas, M., and Hayes, R. Influence of characterised natural organic material on activated carbon adsorption: II. Effect on pore volume distribution and adsorption of 2-methylisoborneol. *Water Res.*, 1997, **31**(5), pp. 1065–1073.

34. Newcombe, G., Morrison, J., and Hepplewhite, C. Simultaneous adsorption of MIB and NOM onto activated carbon. I. Characterisation of the system and NOM adsorption. *Carbon*, 2002, **40**(12), pp. 2135–2146.

35. Kolpin, D. W., Furlong, E. T., Meyer, M. T., Thurman, E. M., Zaugg, S. D., Barber, L. B., Buxton, H. T., Pharmaceuticals, hormones, and other organic wastewater contaminants in U.S. streams, 1999–2000: A national reconnaissance. *Environ. Sci. Technol.*, 2002, **36**(6), pp. 1202–1211.

36. Boyd, G. R., et al. Pharmaceuticals and personal care products (PPCPs) and endocrine disrupting chemicals (EDCs) in stormwater canals and Bayou St. John in New Orleans, Louisiana, USA. *Sci. Total Environ.*, 2004, **333**(1–3), pp. 137–148.

37. Boyd, G. R., et al. Pharmaceuticals and personal care products (PPCPs) in surface and treated waters of Louisiana, USA and Ontario, Canada. *Sci. Total Environ.*, 2003, **311**(1–3), pp. 135–149.

38. Tixier, C., et al. Occurrence and fate of carbamazepine, clofibric acid, diclofenac, ibuprofen, ketoprofen, and naproxen in surface waters. *Environ. Sci. Technol.*, 2003, **37**(6), pp. 1061–1068.

39. Crisp, M. T., Clegg, E. D., and Cooper, R. L. Special Report on Environmental Endocrine Disruptors: An Effect Assessment and Analysis. 1997, U.S. Environmental Protection Agency.

40. Snyder, S. A., Westerhoff, P., Yoon, Y., and Sedlak, D. L. *Pharmaceuticals, Personal Care Products, and Endocrine Disruptors in Water: Implications for the Water Industry*. 2003, Larchmont, NY: Liebert. 21.

41. Focazio, M. J., Kolpin, D. W., Barnes, K. K., Furlong, E. T., Meyer, M. T., Zaugg, S. D., Barber, L. B., and Thurman, M. E. A national reconnaissance for pharmaceuticals and other organic wastewater contaminants in the United States—II) Untreated drinking water sources. *Sci. Total Environ.*, 2008, **402**(2–3), pp. 201–216.

42. Snyder, S. A. Occurrence, treatment, and toxicological relevance of EDCs and pharmaceuticals in water. *Ozone Sci. Eng.*, 2008, **30**(1), pp. 65–69.

43. Redding, A. M., et al. A QSAR-like analysis of the adsorption of endocrine disrupting compounds, pharmaceuticals, and personal care products on modified activated carbons. *Water Res.*, 2009, **43**(15), pp. 3849–3861.

44. Snoeyink, V. L., and Summers, R. S. Adsorption of organic compounds, in *Water Quality and Treatment: A Handbook of Community Water Supplies* (Letterman. R. D., ed.), 1999, Chapter 13, 5th ed., New York: McGraw Hill.

45. Speth, T., and Miltner, R. Technical note: Adsorption capacity of GAC for synthetic organics. *J. Am. Water Works Assoc.*, 1998, **82**, pp. 72–75.

46. Gu, Z., Fang, J., and Deng, B. Preparation and evaluation of GAC-based iron-containing adsorbents for arsenic removal. *Environ. Sci. Technol.*, 2005, **39**(10), pp. 3833–3843.

47. Chen, W., et al. Arsenic removal by iron-modified activated carbon. *Water Res.*, 2007, **41**(9), pp. 1851–1858.

48. Jang, M., Chen, W., and Cannon, F. S. Preloading hydrous ferric oxide into granular activated carbon for arsenic removal. *Environ. Sci. Technol.*, 2008, **42**(9), pp. 3369–3374.

49. Nieto-Delgado, C., and Rangel-Mendez, J.-R. Anchorage of iron hydro(oxide) nanoparticles onto activated carbon to remove As(V) from water. *Water Res.*, 2012, **46**(9), pp. 2973–2982.

50. Parette, R., Cannon, F. S., and Weeks, K. Removing low ppb level perchlorate, RDX, and HMX from groundwater with cetyltrimethylammonium chloride (CTAC) pre-loaded activated carbon. *Water Res.*, 2005, **39**(19), pp. 4683–4692.

51. Parette, R., and Cannon, F. S. The removal of perchlorate from groundwater by activated carbon tailored with cationic surfactants. *Water Res.*, 2005, **39**(16), pp. 4020–4028.

52. Krupa, et al. *GAC: Pore Structure versus Dye Adsorption*. Vol. 88. 1996, Denver, CO: American Water Works Association.

53. Nowack, K. O., Cannon, F. S., and Mazyck, D. W. Enhancing activated carbon adsorption of 2-methylisoborneol: Methane and steam treatments. *Environ. Sci. Technol.*, 2004, **38**(1), pp. 276–284.

54. Olivier, J. P. *The Determination of Surface Energetic Heterogeneity Using Model Isotherms Calculated by Density Functional Theory.* in *Fifth International Conference on the Fundamentals of Adsorption.* 1995. Pacific Grove, California, USA.

55. Seaton, N. A., Walton, J. P. R. B., and Quirke, N., A new analysis method for the determination of the pore size distribution of porous carbons from nitrogen adsorption measurements. *Carbon,* 1989, **27**(6), pp. 853–861.

56. Crittenden, J. C., Parimi, S., Reddy, H. A. Trynoski, J., Hand, D. W., Perram, D. L., and Summers, R. S. *Predicting GAC Performance with Rapid Small-Scale Column Tests.* Vol. 83. 1991, Denver, CO: American Water Works Association. 11.

57. Moore, B. C., Cannon, F. S., Westrick, J. A., Metz, D. H., Shrive, C. A., DeMarco, J., and Hartman, D. J. Changes in GAC pore structure during full-scale water treatment at Cincinnati: A comparison between virgin and thermally reactivated GAC. *Carbon,* 2001, **39**(6), pp. 789–807.

58. Moore, B. C. Comparison of full-scale GAC: Virgin versus reactivated at Cincinnati, Ohio., in *Civil and Environmental Engineering.* 2002, The Pennsylvania State University: University Park.

59. Moore, B. C., Cannon, F. S., Metz, D. H., and Demarco, J. *GAC Pore Structure in Cincinnati Curing Full-Scale Treatment/reactivation.* Vol. 95. 2003, Denver, CO: American Water Works Association. 10.

60. Metz, D. H., De Marco, J., Pohlman, R., Cannon, F. S., and Moore, B. C. Effect of multiple GAC reactivations on disinfection byproduct precursor removal. *Water Supply* 2004, **4**(4), pp. 71–78.

61. Moore, B. C., Wang, Y. J., Cannon, F. S., Metz, D. H., and Westrick, J. Relationships between adsorption mechanisms and pore structure for the adsorption of natural organic matter by virgin and reactivated granular activated carbons during water treatment. *Environ. Eng. Sci.,* 2010, **27**(2), pp. 187–198.

62. Crosson, K. M. *Assessing the Effects of Drinking Water Treatment Processes on Refractory NOM Fractions Using Tetramethylammonium Hydroxide (TMAH)-Thermochemolysis,* PhD thesis, 2005, The Pennsylvania State University: University Park.

63. Nowack, K. O., and Cannon, F. S. Control of calcium buildup in GAC: Effect of iron coagulation. *Carbon,* 1997, **35**(9), pp. 1223–1237.

64. Nowack, K. O., Cannon, F. S., and Arora, H. (1999). *Ferric Chloride Plus GAC for Removing TOC*, Vol. 91, American Water Works Association.

65. Patterson, J. P., Parette, R., and Cannon, F. S. Oxidation of intermediate sulfur species (thiosulfate) by free chlorine to increase the bed life of tailored granular-activated carbon removing perchlorate. *Environ. Eng. Sci.*, 2010, **27**(10), pp. 835–843.

66. Chen, W., Cannon, F. S., and Rangel-Mendez, J. R. Ammonia-tailoring of GAC to enhance perchlorate removal. II: Perchlorate adsorption. *Carbon*, 2005, **43**(3), pp. 581–590.

67. Patterson, J., et al. Competition of anions with perchlorate for exchange sites on cationic surfactant-tailored GAC. *Environ. Eng. Sci.*, 2011, **28**(4), pp. 249–256.

68. Chen, W. F., et al. Arsenic adsorption *via* iron-preloaded activated carbon and zero-valent iron. *J. Am. Water Works Asso.* 2008, **100**(8), pp. 96–105.

69. Chen, W. F., Parette, R., Cannon, F. S., Huang, H., Sheehan, W., Robinson, J., Goye, N., and Smith, R. *Pilot Study on Arsenic Removal with GAC and Zero-Valent Iron*. 2011, The Pennsylvania State University.

70. Jang, M., et al. Combined hydrous ferric oxide and quaternary ammonium surfactant tailoring of granular activated carbon for concurrent arsenate and perchlorate removal. *Water Res.*, 2009, **43**(12), pp. 3133–3143.

Chapter 9

Adsorption of Volatile Organic Vapours

Peter Lodewyckx

Department of Chemistry, Royal Military Academy,
Renaissancelaan 30, Brussels, B-1000, Belgium

Peter.Lodewyckx@rma.ac.be

9.1 Introduction

The term "volatile organic compound", or "VOC", is not really a scientific one, nor is it unambiguous. Generally speaking, VOCs are organic compounds that have a low boiling point, i.e. a high vapour pressure. There have been some efforts to more rigid definitions, e.g. by the European Community [1]. They define an organic compound as "any compound containing at least the element carbon and one or more of hydrogen, oxygen, sulphur, phosphorus, silicon, nitrogen, or a halogen, with the exception of carbon oxides and inorganic carbonates and bicarbonates". It becomes volatile if "having an initial boiling point less than or equal to 250°C measured at a standard pressure of 101.3 kPa".

VOCs can either occur naturally in the environment or can me manmade. Most of them present low to medium solubility in water.

Green Carbon Materials: Advances and Applications
Edited by Thomas E. Rufford, Denisa Hulicova-Jurcakova, John Zhu

ISBN 978-981-4411-13-4 (Hardcover), 978-981-4411-14-1 (eBook)
www.panstanford.com

The main concern with VOCs is the health risks they present. Many of these compounds are toxic to human beings, or to the environment. Several of them are known or suspected to be carcinogenic. Finally, some VOCs contribute to the formation of photochemical pollution, the so-called photochemical smog [2].

The main problem in capturing or eliminating VOCs is precisely their versatility. As their physico-chemical characteristics can differ significantly, technologies based on chemical interactions will only work for a limited range of compounds, e.g. halogenated ones. Other methods to remove VOCs from ambient air, such as absorption, are limited by their poor water solubility. Consequently, the most common method for the elimination of VOCs from air is adsorption onto a suitable adsorbent. Once again, the search for the right adsorbent is dominated by the very large spectrum of these chemical compounds that differ in volatility, polarity, reactivity, size, etc. The adsorbent of choice has to be capable of retaining all, or most of them, rather than having to be tailored for each individual compound. This directs us automatically towards activated carbon.

9.2 Physisorption

9.2.1 Polanyi Theory or TVFM Model

The process of physisorption is the result of a-specific interactions between the carbon surface and the chemical compound: the van der Waals forces. These forces are weak and, as such, present a very limited range. At very close ranges between a carbon atom at the surface and an adsorbate molecule, the repulsion force between the two atomic nuclei becomes more important than the attraction force. This is characterized by the Lennard–Jones potential [3]:

$$E(r) = -C \cdot r^{-6} + B \cdot r^{-12}, \tag{9.1}$$

where r is the distance between the two atoms, i.e. between the adsorbate molecule and the carbon surface. And C and B are empirical constants, respectively, related to the attraction and repulsion forces.

In very small pores, with distances between the walls that are of the same order of magnitude as the diameter of the adsorbate molecule, the potentials from the opposing walls will overlap (see Fig. 9.1).

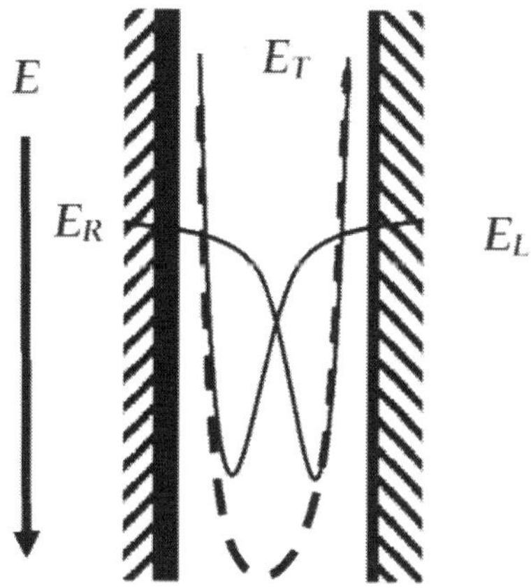

Figure 9.1 Lennard-Jones potentials of L (left) and R (right) walls of a micropore and the resulting T (total) potential as "felt" by an adsorbate molecule.

The resulting overall potential sink is very strong, resulting in an attraction force that is much higher than the normal van der Waals force between a single wall and a molecule. The adsorbate is literary "trapped" in the small pore. According to this theory smaller pores (i.e. higher overlap potentials) will be filled with adsorbate molecules before adsorption starts in bigger ones. On a macroscopic scale it looks as if the pore system is getting gradually (from small pores to larger ones) filled with molecules. As these molecules are just stacked upon each other in the pores like they are "poured in" as a liquid, this theory, proposed by Polanyi, is also called the theory of volume filling of micropores (TVFM) [3]. This theory is also confirmed by the fact that when calculating the amount of vapour adsorbed, one finds the density of the adsorbed phase to be very close to the liquid density, provided the temperature at which the isotherm was taken is well below the critical temperature. Several authors have found deviations to this rule [4–6], but only in very narrow pores where, by physical space constraints, a more ordered structure has to exist.

Once the pore becomes so big its walls are too far apart for their potentials to overlap, the volume filling stops and the adsorption mechanism reverts to the normal interactions between the adsorbate molecules and the carbon pore walls. It is clear that the size of the pores at which the volume filling stops is depending on the size of the adsorbate molecule and the strength of the interactions with the pore walls. This is not necessarily exactly at the conventional maximum size of the micropores (20 Å according to the IUPAC classification) [3]. Sometimes the volume filling stops at smaller sizes, but more often it will extend to sizes between 20 and 30 Å. This can be graphically shown by the DR model (see Section 9.2.2.1).

9.2.2 Isotherm Models and Micropore Volume

Isotherms are often used to characterise adsorbents. These are graphs representing the amount adsorbed at equilibrium versus the partial pressure of the vapour in the atmosphere at a certain temperature. However, in itself an isotherm also represents the adsorption process. Inversely, isotherms for a certain vapour can be derived from the pore size distribution of a carbon that has been obtained from an isotherm for another vapour or by other means such as small angle x-ray scattering (SAXS) or transmission electronic microscopy (TEM). There exist several models to describe the isotherm. Most of them are based on theoretical assumptions on the physics and chemistry of the adsorption process. As adsorption in the micropore system can be adequately described by the TVFM model, the isotherm equation based on this model, as proposed by Dubinin and co-workers [7], is widely used.

9.2.2.1 Dubinin plots

According to Dubinin, the force involved in the TVFM is $-\Delta G$, the differential free energy of adsorption (also called "adsorption potential" by Polanyi). It is defined by $-\Delta G = R \cdot T \cdot \ln(p_0/p)$. The basic postulate of the Dubinin theory is therefore the relationship between θ, the degree of filling of the micropores, and $-\Delta G$:$\theta = \varphi(-\Delta G/\beta)$. The parameter β is a scaling factor to take into account the differences between the adsorptives; conventionally β is taken to be equal to 1 for benzene. This is the second assumption made by Dubinin: It implies the ratio $-\Delta G_1/-\Delta G_2$ for two given adsorbates to be independent from the degree of pore filling, θ. This means that according to this theory, the mechanism of adsorption and the energy involved do not change over the range of values of θ for which the model is valid. The third assumption of the theory, as proposed by Dubinin and Radushkevich, is the most disputed one: It assumes a Gaussian pore size distribution. While this condition is seldom fulfilled in reality, the resulting equation gives very good fits for the isotherms on activated carbon:

$$W = W_o \exp\left[-\frac{BT^2}{\beta^2}\log^2\left(\frac{p_o}{p}\right)\right], \qquad (9.2)$$

where W is the volume of pores filled at p/p_o (cm^3/g), W_o the total volume of the micropore system (cm^3/g), B a structural constant related to the width of the Gaussian pore distribution (K^{-2}), T the temperature at which the isotherm has been taken (K), β a similarity constant, depending solely on the adsorbate ($\beta = 1$ for benzene)(–) and p_o/p the inverse of the relative pressure of the adsorbate (–).

For the adsorption of gases and vapours, one does not measure W, but the volume adsorbed from the gas phase. In order to calculate W, one has to know the density of the adsorbate in its adsorbed state. As one assumes a mechanism of pore filling, this density ρ^* can be taken as equal to the liquid density of the adsorbate d_L. The curve of $\ln(W)$ versus $\log^2(p_o/p)$ gives a straight line in the domain of validity of the DR equation. The Y-intercept equals the logarithm of the micropore volume W_o. The slope of the line is proportional to B/β and, once β has been calculated [8, 9], to B, which, in turn, is related to the characteristic adsorption energy of the carbon E_o.

Normally deviations from the straight line can be found for very low and very high values of $\log^2 (p_o/p)$. The first corresponds to values of p/p_o were the adsorption shifts from micropores to mesopores. This results in an upward swing of the DR curve because of the excess adsorption which is not accounted for by the pore-filling theory [10]. The second is due to either activated diffusion or molecular sieving effects [11].

Several generalisations have been proposed for the DR equation. One of them is the Dubinin–Astakhov or DA equation [3]. It has the same base as the DR equation except that it assumes a Weibull rather than a Gaussian distribution of the micropores, resulting in an adjustable exponent "*n*" rather than a fixed value of "2" as for the DR equation. Several authors have proposed further generalisations by expressing W as a summation of two to n [12–13] DR equations, one for each individual pore group of a given size. The application of this equation is rather tedious and is not used very often, in spite of its inherent logic. For most isotherms of organic vapours on commercial activated carbons very good fits are obtained with the basic (DR) equation.

Usually either N_2 or Ar at low temperatures (77–87 K) or CO_2 at higher temperatures (273–293 K) are used as "probe" molecules to assess the porosity of carbons, i.e. to calculate W_0 and E_0 (or B), which are needed to calculate the adsorption capacity for VOCs.

9.2.2.2 Other isotherm models

Another very common way to describe the adsorption into the (micro)pore system is the density functional theory (DFT). This theory is based on the minimisation of the grand potential function called, $\Omega_v[\rho]$, with respect to the density $\rho(z)$ to obtain the equilibrium density profile [14]. The parameter z being the co-ordinate normal to the surface in the case of slit shaped pores. We will not get into the mathematics of this model, but the resulting equations can only be solved numerically and even then only by the use of supercomputers. To overcome this problem, the experimental isotherm will be compared with previously calculated model isotherms: each pore size will contribute to the total experimental isotherm in proportion to the fraction of the total pore volume of the sample it represents. The solution to this equation requires a set of previously calculated model isotherms for each pore size with the same adsorbate on a similar type of adsorbent. Then it can be solved numerically (e.g. least squares method) to give the most probable set of values for the pore size distribution. The main advantage of this method is its validity for both micro- and mesopores. There are, however, some major setbacks: The deconvolution is not always straightforward and false minima can appear as mathematical artifacts. The biggest disadvantage is the lack of model isotherms for organic vapours. The existing models focus on conventional probes for physisorption capacity such as N_2 at 77 K, CO_2 at 273 K and Ar at 87 K. The maximum uptake of an organic compound can be calculated by taking the micropore volume, derived from, for example, a N_2 isotherm, and assuming a complete filling at liquid density. However, a partial uptake cannot be calculated without the knowledge of the model isotherms for that particular organic vapour.

Other isotherm models such as Langmuir, Freundlich, and Brunauer–Emmett–Teller (BET) can fit either partially or completely isotherms from both VOCs and conventional probe gases (N_2, CO_2, Ar) [3]. They lack, however, the physical consistency of the DR and DFT approaches and are therefore less reliable in predicting the correct total micropore volume and the ensuing physisorption capacity.

9.2.3 Physisorption Capacity

It appears that for volatile organic compounds, the potential in the mesopores (i.e. without an overlap) is too weak to attain any adsorption. Hence, the theoretical maximum capacity of an activated carbon to adsorb a VOC is its total micropore volume. In order to calculate this capacity on a weight basis, one must simply use the liquid density of the adsorbate. This is also known as the Gurvich rule. In order to know how much vapour will be adsorbed at a certain vapour concentration in air one can easily use the Dubinin–Radushkevich equation to calculate W_e, as W_0 and E_0 can be derived from an isotherm with a probe molecule (usually N_2) and β can be estimated by different models [8, 9].

9.2.4 Selectivity

It is clear that some vapours are more easily adsorbed than others (due to the influence of similarity coefficient β in the DR equation). This will have an influence on the co-adsorption of different vapours. There also exists the possibility of a molecular sieving effect: When the adsorbate molecule becomes too big, it will not enter the smallest pores. This effect has very limited influence in the case of VOC adsorption: Volatility and molecular weight being roughly inversely proportional, most VOCs will be rather small molecules.

9.3 Specific Interactions

9.3.1 Deviations from the TVFM Model

The volume filling of the micropores is the most important VOC adsorption mechanism, but not the only one. In certain cases, there exist specific interactions between the adsorbate molecule and heteroatoms on the carbon surface. These active sites consist usually of oxygen (present at most carbon surfaces as acid carboxylic, lactonic or phenolic groups or as basic groups), nitrogen or sulphur. In some cases, heteroatoms or metal impregnations are artificially introduced in the carbon matrix to enhance specific interactions with certain highly volatile gases. In any case, the surface concentration of these non-carbon atoms is very small. So it is clear that specific interactions will only play a predominant role when the VOC is present in trace concentrations: At higher concentrations, they will not be discernible from the van der Waals interactions and the "normal" volume filling of the micropores will prevail. Another factor is the nature of the specific interaction. Apart from very specific cases (e.g. radicalization of a VOC molecule in the presence of adsorbed water) the interaction will not involve actual chemical bonding (ionic or covalent) but only hydrogen bridging or analogous δ^{+}-δ^{-}-interactions. This is only possible when the molecule presents a high local polar behaviour, i.e. for small, polar molecules such as short alcohols.

Combining both aspects, we can conclude that specific interactions are limited to the adsorption of trace quantities of small, polar VOCs [15–18].

9.3.2 Isotherm Models

In these cases, the first part of the isotherm (i.e. low VOC concentration) cannot be fitted with a DR equation. Usually Langmuir (see Eq. 9.3) or Freundlich [3] equations are used, even if these were originally developed to describe adsorption on non-porous, flat surfaces:

$$n = \frac{n_m \cdot b \cdot (p/p_o)}{\left[1 + b \cdot (p/p_o)\right]} \tag{9.3}$$

In this equation, n represents the amount adsorbed, n_m traditionally the monolayer capacity, b is related to the interaction energy and p/p_0 is the partial VOC pressure. By extension, n_m and b can be used to represent the amount of active sites and how strong they interact with the specific VOC (e.g. a small alcohol chain).

9.4 Adsorption Kinetics

Generally speaking, the adsorption kinetics of VOCs are determined by the slowest step(s) in the adsorption process. This can be the diffusion of the vapour from the bulk phase till the surface of the carbon particle (also called interparticle diffusion), Knudsen diffusion in the mesopores, surface diffusion in the micropore system or the actual adsorption process. Recent studies seem to indicate that Knudsen diffusion is never the rate controlling step in the adsorption of a pure VOC [19]. In most cases, the kinetics seem to be governed by a combination of bulk and surface diffusion [20–22].

It is logical to presume that at very low concentrations, and a high micropore surface area, bulk and surface diffusion, respectively, become very fast and the actual adsorption could become the rate controlling step. At higher concentrations, even if some part of the actual adsorption is still specific, the kinetics will be governed by the transport to the active sites and one observes a similar behaviour and similar rate constants as in the case of a pure physisorption. There is little or no research in this area, although the effect of the rate controlling step being dependent on the inlet concentration has indeed been observed in the adsorption of small inorganic molecules [23].

9.5 Filter Breakthrough Time

In most cases, the adsorption of VOCs is part of a filtration process destined at removing them from the ambient air. As such, one is particularly interested by the breakthrough time. This is the time between the first moment the filter is solicited by the contaminated air stream and the moment one can detect a certain, predetermined, concentration of the contaminant at the end of the filter. This concentration is dependent on the application and the VOC itself, especially its toxicity. Usually it is taken lower than or equal to the toxic threshold of the vapour, i.e. at a fraction (1% or less) of the concentration in the air stream.

9.5.1 Modelling Filter Breakthrough

The broad range of toxic agents, environmental conditions and types of activated carbon simply does not allow laboratory testing of each and every possible combination. In order to evaluate the protection provided by a specific activated carbon filter in a well-defined environment, it is necessary to trust the results of mathematical modelling, taking into account the necessary safety margins.

9.5.1.1 Different types of models

The chemical engineering models consider the filter as a chemical reactor, using the conventional equations for mass and heat transfer to describe the transport processes through the filter [22, 24–26]. Even with modern computational techniques, one has to make some assumptions and simplifications to the real phenomena. Usually the following set of equations is used:

- a mass balance over the filter (interparticle diffusion)
- an energy (heat) balance over the filter
- a mass balance over one carbon particle (intraparticle diffusion)
- an energy (heat) balance over one carbon particle

As one can see, this is a system of coupled differential equations. Some of these balances can even involve more than one equation, e.g. intraparticle diffusion that can be a combination of Knudsen diffusion and surface diffusion. A supplementary equation is needed to describe the adsorption itself. In the case of pure physisorption, this is simply the mathematical description of the adsorption isotherm, e.g. the Dubinin–Radushkevich equation. A combination of these equations, including a correct set of boundary conditions, will yield a time-concentration profile through the filter.

These models present two significant advantages: They simulate the complete breakthrough curve and they are, within the limits of the assumptions and simplifications, physically correct. The complete breakthrough curve is of particular interest in gas separation applications and more complex systems such as PTSA (pressure and temperature swing adsorption).

The main advantage, i.e. the physical correctness, of these models comes at a price as one has to describe, in detail, every physical and chemical phenomenon inside the carbon and one has to know all the interaction parameters. Some of these, such as the tortuosity, the different diffusion coefficients, the (micro-) flow pattern and chemical interactions, are difficult to calculate or even to evaluate from a limited number of experiments.

9.5.1.2 Wheeler–Jonas model

If we just consider the protection provided against VOCs, it is sufficient to know the breakthrough time. Several models exist, with different degrees of complexity [27]. One of the most widely used is the Wheeler–Jonas (WJ) model [28]. This model calculates the breakthrough time as a combination of two terms: the capacity of the filter (assuming infinitely fast adsorption kinetics) and the broadness of the adsorption front, i.e. the axial dispersion of the adsorption front due to finite diffusion constants. The prediction is based solely on measurable and readily available macroscopic parameters:

$$t_b = \frac{M \cdot W_e}{Q \cdot c_{in}} - \frac{W_e \cdot \rho_b}{k_v \cdot c_{in}} \ln\left[\frac{c_{in} - c_{out}}{c_{out}}\right], \tag{9.4}$$

where t_b is the breakthrough time to reach c_{out} (min), c_{in} the vapour inlet concentration in air (g/cm^3), c_{out} the chosen breakthrough concentration (g/cm^3), M the weight of the carbon bed (g), W_e the equilibrium adsorption capacity (g/g$_{carbon}$), Q the volumetric flow rate (cm^3 /min), ρ_b the bulk density of the carbon bed (g$_{carbon}$/cm^3) and k_v the overall adsorption rate coefficient (min^{-1}). Two unknown parameters in this equation need to be calculated: W_e and k_v. Several models have been proposed to calculate them without any prior breakthrough experiment. The capacity W_e is normally estimated directly from the DR equation [9]. Calculation of the overall adsorption rate coefficient k_v is more complex. For organic compounds, the kinetics seems to be governed by a combination of bulk and surface diffusion. The most recent and complete semi-empirical equation to calculate the value of k_v is Eq. (9.5) [21]

$$k_v = 800 \cdot \beta^{0.33} \cdot v_L^{0.75} \cdot d_p^{-1.5} \cdot \sqrt{\left(\frac{W_e}{M_w}\right)}, \tag{9.5}$$

where k_v is the overall adsorption rate coefficient (min^{-1}), W_e the adsorption equilibrium capacity (g/g_{carbon}), M_w the molecular weight of the vapour (g/mole), d_p the average diameter of the carbon particle (cm), β the affinity coefficient of the organic vapour and v_L the linear velocity through the bed (cm/s). Even if it is not apparent, this equation strongly depends on temperature. Not only through W_e (see Eq. 9.2), but some preliminary results seem to indicate that the pre-exponential value of 800 is in fact strongly dependent ($\sim T^4$) on temperature.

9.6 Environmental Factors Influencing Adsorption

9.6.1 Temperature

Temperature has two, diverging, influences on the adsorption and breakthrough time. As one can see from the DR equation (Eq. 9.2), higher temperatures will result in a lower adsorption capacity. On the contrary, they will slightly increase adsorption kinetics as W_e is a negative function of temperature, but the pre-exponential term a positive one. Given the fact that in most applications the first (capacity) term of the WJ equation is three to five times higher than the second (kinetic) one [29], the overall impact of temperature on the breakthrough time will usually be a negative one. Especially in the case of very fast adsorption kinetics (see Section 9.7.2), the negative effect on capacity will largely outweigh the small beneficial one on kinetics.

9.6.2 Humidity

Activated carbon is known to be hydrophobic. A discussion on the water vapour–activated carbon interactions, and the resulting isotherms, lies outside the scope of this chapter. In short, activated carbon only starts adsorbing water vapour at high relative concentrations. However, in most applications, the amount of water vapour present in the air is very large (more than 60%

relative humidity). At these high quantities, activated carbon does adsorb water quite well and there will be interactions between the (adsorbed) water vapour and the VOC. Most VOCs are not soluble in water. Therefore, the influence of water vapour on their breakthrough time can be estimated by so-called exclusion models. These models [29–31] consider the available micropore volume to be divided between the water and the VOC. This partition of the micropore volume, and the resulting loss in adsorption capacity for the VOC, can be calculated on the basis of the respective concentrations present in the air stream and the affinity of the carbon for the VOC. As a rule of thumb, more volatile VOCs will be more (negatively) influenced by the humidity of the environment. Not only the capacity but also the adsorption kinetics will suffer. Generally speaking, a higher relative humidity will result in slower VOC adsorption kinetics, especially for carbons with a limited mesopore volume.

There is very little research on the influence of relative humidity on water-soluble organic vapours, especially because they are fewer in numbers than insoluble VOCs. However, there is a general understanding they suffer less from the adverse effects described above [32, 33].

9.6.3 Other Vapours or Gases

In a real environment, it is not only water vapour that causes an interference with VOC adsorption. In many applications, more than one VOC is present in the atmosphere (e.g. laboratories, sewage plant, landfill, agricultural applications). This means the different VOCs will compete for the available micropore volume. As in most cases humidity is also present, the interactions become very complex [34–40]. One of the typical effects of multicomponent adsorption is the so-called roll-up effect. This can be illustrated by the case of two vapours. The affinity of the carbon for the two vapours being different, one of them will be preferably adsorbed. This means that there will be co-adsorption till all micropores are filled, but then the more strongly adsorbed vapour will start replacing the weaker adsorbed one. As a result, this weaker adsorbed one will be pushed away by the adsorption front of the latter, resulting in some kind of tidal wave. Finally, the weaker adsorbed one will not only break through the filter much earlier than when applied as a single vapour, but its concentration will also rise above the inlet concentration as a result of the amount of vapour assembled in the wave. This is shown in Fig. 9.2 for a typical breakthrough curve of a bi-component adsorption. The roll-up effect can be visualized by x-ray microtomography [40].

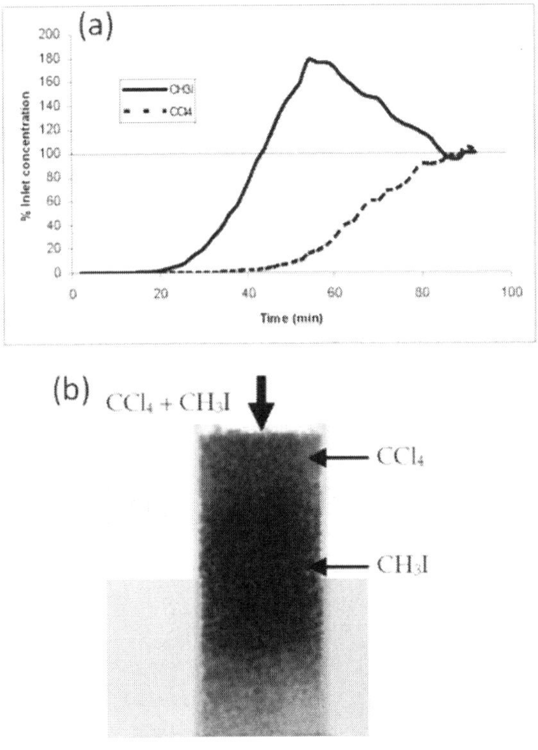

Figure 9.2 Roll-up effect: (a) Breakthrough curve; (b) X-ray microtomographic image of simultaneous adsorption of CCl_4 (= stronger) and CH_3I (= weaker adsorbed).

9.7 Influence of the Physical Form of the Carbon

9.7.1 Granular Carbon

The physical form of activated carbon is usually dictated by the specific application and choices will be governed by researching the optimum combination of capacity, kinetics, pressure drop and price. Granular activated carbon (GAC) is by far still the most used form of activated carbon. Granules can be irregular in form or can be extrudates (see Fig. 9.3) typically with diameters between 0.1 and 1 cm. The main advantages of GAC are its low cost and its good trade-off between capacity and pressure drop. The first one is due to the low cost of the raw materials (valid for most types of activated carbon) and of the production mechanism. The latter is very important for air filtration: larger carbon beds have a higher capacity but present a higher pressure drop. This leads to in an increased breathing resistance in case of gas mask filters or extra energy consumption in case of assisted filtration (e.g. ventilation).

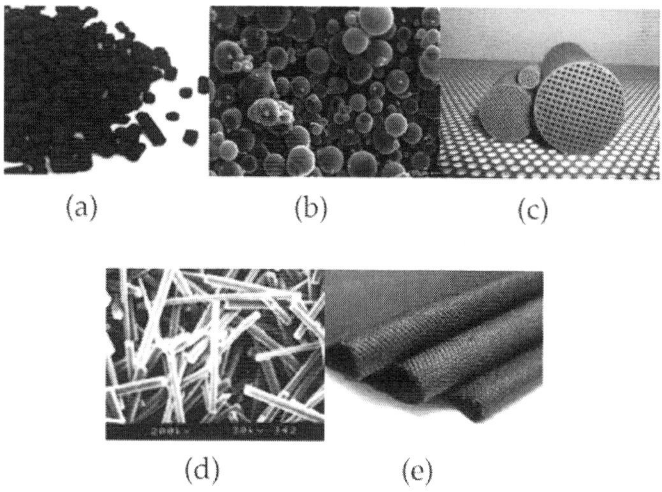

Figure 9.3 Different forms of activated carbon: (a) Granular, (b) microbeads, (c) monoliths, (d) fibres, (e) cloth.

9.7.2 Powdered Carbon

Powdered carbons are well known in the area of water purification. The fine powder is dispersed into the water where the small particles are responsible for fast adsorption kinetics. They are, however, seldom used in the gas phase removal of VOCs as it is highly unpractical and inefficient to inject them in an air stream. Filling a filter with powdered carbon would lead to an unacceptable pressure drop even for very shallow beds.

9.7.3 Carbon Microbeads

Carbon microbeads represent a good trade-off between capacity, kinetics and pressure drop. These beads typically have diameters around 250–500 μm (Fig. 9.3). This means the adsorption rate is very high, i.e. the second term of the WJ equation becomes nearly negligible, without inacceptable pressure drops for shallow beds. They are, however, not very well suited for larger beds as the pressure drop is still more important than in the case of GAC.

9.7.4 Carbon Monoliths

Activated carbon monoliths (ACM) have been around for some time but have attracted a lot of interest lately. The contaminated air flows through a number of channels (Fig. 9.3) and is adsorbed in the micropores present in the walls. The monolith itself can be made entirely of carbon, or can be a support (e.g. ceramic) coated with a carbon layer [41]. The main advantage lies in the nearly negligible pressure drop, even for very long monoliths. However, they are also known for rather fast kinetics; it has been suggested that the rate of adsorption is proportional to the width of the channels (instead of to the size of the granules as it is the case for GAC) [42]. A last, but very important, plus is their regenerability. Resistive heating of the monolith support allows for a good temperature based desorption of the VOCs, at low levels of energy consumption.

Activated carbon monoliths do have some disadvantages. The first is the capacity. Even though some ACM present very high capacities, they will always be slightly lower than GAC on a weight or a volume basis as not the whole structure is used as an adsorbent. Another area of concern is leakage. Even though certain studies seem to indicate that leakage is not an issue [43], this is contradicted by others [41]. It is indeed not impossible that certain gas molecules will pass through the channels of the monolith without getting into contact with the channel walls, i.e. without being adsorbed. This number of molecules is evidently very low, and inversely proportional with the length of the monolith. As a result, leakage can be ignored in most cases, except in applications involving highly toxic VOCs. In this case supplementary precautions have to be taken. Finally, even though still cheap compared to non-carbon adsorbents, ACM are clearly more expensive than GAC due to the more elaborate production process.

9.7.5 Carbon Fibres and Carbon Cloth

Activated carbon fibres (ACF; see Fig. 9.3d) can be used as such, or in woven or non-woven form (felt, cloth). Their biggest advantage is the fast kinetics, comparable to the microbeads. The combination of very low diameters and the micropores being situated at the surface of the fibres results indeed in very high k_v values for the WJ equation [44]. Another advantage is their physical form. Theoretically, it is possible to integrate the flexible ACF into other equipment (e.g. clothing) in any desired shape. At this time however, only a limited number of such applications are available. This is not only due to the price tag but also due to the fact that ACF have a problem with the trade-off between capacity and pressure drop: Pressing the fibres together or piling up several sheets of cloth enhances the capacity per unit of volume but increases dramatically the pressure drop.

9.7.6 Carbon Nanotubes

Carbon nanotubes and other novel forms of carbon (single- and multi-walled nanotubes and nanohorns) can present very interesting adsorption properties. However, due to the elaborate production techniques and the associated cost, they are not (yet) used in real life VOC adsorption applications.

9.7.7 Combining Different Physical Forms

As stated earlier, the choice of the physical form of the adsorbent will be the result of a matching between its advantages and disadvantages and the specific needs of the application. Currently there exists a tendency towards obtaining the optimum solution by a combination of different physical forms. For example, one can take the case of a regenerable, protective filter against high concentrations of a very toxic VOC in a high velocity air stream. The need for regeneration leads towards monoliths. So does the high velocity, as the pressure drop is usually assumed to be proportional to the square of the linear velocity (e.g. D'Arcy–Weisbach equation).

However, as the leakage risk cannot be excluded, the use of monoliths seems to be undesirable. In this case, the solution could exist of an activated carbon monolith, followed by one or two layers of activated carbon cloth. This way the evident advantages of the monolith are retained whilst the cloth provides for the fast capture of any VOC molecules leaking through. This possible leakage being small, the limited amount of cloth covers the capacity requirement without adding any significant pressure drop.

9.8 Conclusions

VOCs constitute a major problem in our modern society. Adsorption is one of the most cost-effective ways of dealing with this problem. Activated carbon is certainly one of the best-suited adsorbents as in most cases there is no need for specific interactions in order to obtain adsorption. As such, activated microporous carbons will be efficient by their non-specific adsorption behaviour and their very large available micropore volume. Activated carbon is also available under different physical forms that allow a good tailoring for a specific application.

References

1. Directive 2004/42/CE of the European Parliament and of the Council of 21 April 2004 on the limitation of emissions of volatile organic compounds due to the use of organic solvents in certain paints and varnishes and vehicle refinishing products, *Official J. Eur. Union* 30.4.2004.
2. Seinfeld, J. (1986). *Atmospheric Chemistry and Physics of Air Pollution*, 1st ed. (John Wiley and Sons, Inc., Somerset NJ, USA).
3. Gregg, S. J., Sing, K. S. W. (1982). *Adsorption, Surface Area and Porosity*, 2nd ed. (Academic Press Inc. (London) Ltd., London, UK).
4. Iiyama, T., Nishikawa, K., Otowa, T., Kaneko, K. (1995). An ordered water molecular assembly structure in a slit-shaped carbon nanospace, *J. Phys. Chem.*, **99**, pp. 10075–10076.
5. Dubinin, M. M., Neimark, A. V., Serpinsky, V. V. (1993). Impact of the adsorbate compressibility on the calculation of the micropore volume, *Carbon*, **31**, pp. 1015–1018.

6. Aukett, P. N., Quirke, N., Riddiford, S., Tennison, S. R. (1992). Methane adsorption on microporous carbons—a comparison of experiment, theory, and simulation, *Carbon*, **30**, pp. 913–924.

7. Dubinin, M. M. (1967). Adsorption in micropores, *J. Colloids Interface Sci.*, **23**, pp. 487–499.

8. Kenneth, E. Noll, Dahui Wang, Ting Shen (1989). Comparison of three methods to predict adsorption isotherms for organic vapors from similar polarity and nonsimilar polarity reference vapors, *Carbon*, **27**, pp. 239–245.

9. Wood, G. O. (1992). Activated carbon adsorption capacities for vapors, *Carbon*, **30**, pp. 593–599.

10. Carrot, P. J. M., Roberts, R. A., Sing, K. S. W. (1987). Adsorption of nitrogen by porous and non-porous carbons, *Carbon*, **25**, pp. 59–68.

11. Rodriguez-Reinoso, F. Torregrosa, R. (1990). The analysis of narrow microporosity in activated carbons, *Carbon Conference 1990, Paris, France*, pp. 26–27.

12. Dubinin M. M., Stoeckli, H. F. (1980). Homogeneous and heterogeneous micropore structures in carbonaceous adsorbents, *J. Colloids Interface Sci.*, **75**, pp. 34–42.

13. Kowalczyk, P., Terzyk, A. P., Gauden, P. A. (2002). The application of a CONTIN package for the evaluation of micropore size distribution functions, *Langmuir*, **18**, pp. 5406–5413.

14. Olivier, J. P. (1995). Modeling physical adsorption on porous and nonporous solids using density functional theory, *J. Porous Mater.*, **2**, pp. 9–17.

15. Bradley R. H. R. B. (1991). A comparison of the adsorption behaviour of nitrogen, alcohols, and water towards active carbons, *Carbon*, **29**, pp. 1165–1172.

16. Lopez-Ramon, M. V., Stoeckli, F., Moreno-Castilla, C., Carrasco-Marin, F. (2000). Specific and non-specific interactions between methanol and ethanol and active carbons, *Langmuir*, **16**, pp. 5967–5972.

17. Salame, I., Bandosz, T. (2000). Adsorption of water and methanol on micro- and mesoporous wood-based activated carbons, *Langmuir*, **16**, pp. 5435–5440.

18. Muller, E. A., Hung, F. R., Gubbins, K. E. (2000). Adsorption of water vapor—methane mixtures on activated carbons, *Langmuir*, **16**, pp. 5418–5424.

19. Lodewyckx, P., Berezovska, I., Rodriguez-Reinoso, F., Juarez-Galan, J. M., Silvestre-Albero, A. (2011). The influence of mesopores on the kinetics of organic vapour adsorption by activated carbons, *CD proceedings of the International Carbon Conference, Shanghai, China*.

20. Busmundrud, O. (1993). Vapour breakthrough in activated carbon beds, *Carbon*, **31**, pp. 279–286.

21. Wood, G. O., Lodewyckx, P. (2003). An extended equation for rate coefficients for adsorption of organic vapors and gases on activated carbons in air-purifying respirator cartridges, *Am. Ind. Hyg. Assoc. J.*, **64**, pp. 646–650.

22. Lodewyckx, P. (2006). Modeling gas phase adsorption in industrial and military application, in *NATO Science through Security Series: C—Environmental Security: Combined and hybrid adsorbents* (Loureiro, J. M., Kartel, M. T. eds.), (Springer).

23. Lodewyckx, P., Verhoeven, L. (2003). Using the modified Wheeler–Jonas equation to describe the adsorption of inorganic molecules: chlorine, *Carbon*, **41**, pp. 1215–1219.

24. Ribeiro, A. M., Loureiro, J. M. (2002). Simulation of toxic gases and vapours removal by activated carbon filters, *Chem. Eng. Sci.*, **57**, pp. 1621–1626.

25. Das, D., Gaur, V., Verma, N. (2004). Removal of organic compound by activated carbon fiber, *Carbon*, **42**, pp. 2949–4962.

26. Linders, M. J. G., Mallens, E. P. J., Bokhoven, J. J. G. M., Kapteijn, F., Moulijn, J. A. (2003). Breakthrough of shallow activated carbon beds under constant and pulsating flow, *Am. Ind. Hyg. Assoc. J.*, **64**, pp. 173–180.

27. Balieu, E. (1976). Characterization of respirator adsorbent filters by means of penetration curve parameters, *Ann. Occup. Hyg.*, **19**, pp. 203–213.

28. Jonas, L., Rehrmann, J. (1972). The kinetics of adsorption of organo-phosphorus vapors from air mixtures by activated carbons, *Carbon*, **10**, pp. 657–663.

29. Lodewyckx, P., Vansant, E. F. (1999). Influence of humidity on adsorption capacity from the Wheeler–Jonas model for prediction of breakthrough times of water immiscible organic vapors on carbon beds, *Am. Ind. Hyg. Assoc. J.*, **60**, pp. 612–617.

30. Marban, G., Fuertes, A. (2004). Co-adsorption of n-butane/water vapour mixtures on activated carbon fibre-based monoliths, *Carbon*, **42**, pp. 71–81.

31. Stoeckli, F. (1998). Recent developments in Dubinin's theory, *Carbon*, **36**, pp. 363–368.

32. Russell, B., LeVan, M. (1997). Coadsorption of organic compounds and water vapor on BPL activated carbon. 3. Ethane, propane, and mixing rules, *Ind. Eng. Chem. Res.*, **36**, pp. 2380–2389.

33. Ustinov, E. A., Polyakov, N. S., Gubkina, M. L. (1998). Dynamics of joint adsorption of mutually soluble substances by activated carbons, *Russian Chem. Bull.*, **47**, pp. 1446–1450.

34. Balzi, M. W., Liapis, A. I., Rippin, D. W. T. (1978). Application of mathematical modeling to the simulation of multi-component adsorption in activated carbon columns, *Trans IChemE*, **56**, pp. 145–156.

35. Morbidell, M., Servida, A., Storti, G., Carra, S. (1982). Simulation of multicomponent adsorption beds. Model analysis and numerical solution, *Ind. Eng. Chem. Fundam.*, **21**, pp. 123–131.

36. Do, D. D., Hu X., Mayfield, P. L. J. (1991). Multicomponent adsorption of ethane, n-butane and n-pentane in activated carbon, *Gas Sep. Purify.*, **5**, pp. 35–48.

37. Hu, X., Rao, G. N., Do, D. D. (1993). A mathematical model for multicomponent adsorption, desorption and displacement kinetics of ethane, propane and n-butane on activated carbon, *Gas Sep. Purif.*, **7**, pp. 197–206.

38. Qiao, S., Wang, K., Hu, X. (2000). Using local IAST with micropore size distribution to predict multicomponent adsorption equilibrium of gases in activated carbon, *Langmuir*, **16**, pp. 1292–1298.

39. Lavanchy, A., Stockli, M., Stoeckli, F. (1996). Binary adsorption on active carbons described by the combined Myers-Prausnitz and Dubinin theories, *Proceedings European Carbon Conference, Newcastle, UK.*

40. Berezovska, I., Verdin, E., Lodewyckx, P. (2011). The use of x-ray tomography for the visualisation of dynamic adsorption of binary organic mixtures, *CD Proceedings of the International Carbon Conference, Shanghai, China.*

41. Valdes-Solis, T., Linders, M., Kapteijn, F., Marban, G., Fuertes, A. (2003). Modelling of the breakthrough performance of carbon-ceramic monoliths in gas adsorption, *CD Proceedings of the International Carbon Conference, Oviedo, Spain.*

42. Lodewyckx, P., Valdés-Solis, T., Linders, M. J. G., Kapteijn, F. (2004). Application of the Wheeler-Jonas equation for the calculation of carbon monolith breakthrough times, *CD Proceedings of the International Carbon Conference, Providence RI, USA.*

43. Tennison, S. (2005). MAST Carbon Ltd: The future for nano-porous carbon materials, *Britisch Carbon Group Meeting, Brighton, March 2005.*

44. Lodewyckx, P., Ryu, S. K. (2002). Modelisation of the adsorption of organic vapours on activated carbon fibres by means of the reaction kinetic equation, *CD proceedings of the International Carbon Conference, Beijing, China.*

Chapter 10

Carbon Foam as a Radionuclide Trap Material for Fast Nuclear Reactors

Prasanta Jana and V. Ganesan

Materials Chemistry Division, Indira Gandhi Centre for Atomic Research, Kalpakkam 603 102, Tamil Nadu, India

pjanaigcar137@gmail.com, ganesh@igcar.gov.in

Carbon foam is a three-dimensional carbonaceous material. It has several applications such as porous electrodes, high-temperature insulators, filters and demisters, storage batteries, scaffolds, and acoustic control medium. It could be synthesized from various precursors such as phenol formaldehyde foam, urethane polymer, cork, and mesophase pitch. This chapter deals with the synthesis of carbon foam from sucrose, an agricultural product and its application as a radionuclide trap material for use in a sodium-cooled fast reactor.

10.1 Carbon Foam

Carbon foams were first developed by researchers in the late 1960s as reticulated vitreous (glassy) carbon. Ford [1] first reported carbon foams produced by carbonizing organic polymer foams (phenol

Green Carbon Materials: Advances and Applications
Edited by Thomas E. Rufford, Denisa Hulicova-Jurcakova, John Zhu

ISBN 978-981-4411-13-4 (Hardcover), 978-981-4411-14-1 (eBook)
www.panstanford.com

formaldehyde foam) through a simple heat treatment process. Googin et al. [2] reported the process dedicated to control the structure and material properties of carbon and graphitic foams by varying the precursor material (partially cured urethane polymer). During the several decades following these initial discoveries, many researchers explored a variety of applications for these materials ranging from electrodes to insulating liners for high-temperature applications up to 2773 K. It can be seen that reticulated carbon foams have been used as template for the preparation of many of the metallic and ceramic foams currently used in industry. In the 1970s, research focused primarily on producing carbon foams from alternate precursors. For example, Klett [3] produced carbon foam from cork, a natural cellular precursor. Others worked on various processes and precursors to modify the properties and to reduce the cost. A rapid method for forming vitreous carbon foams from flexible polyurethane foams was reported by Vinton et al. [4]. An important step in the method is to permit rapid carbonization without cracking.

Carbon foams have been classified as non-graphitic and graphitic. Non-graphitic foams having low thermal and electrical conductivities were prepared by carbonizing organic polymers through heat treatment. Precursors for non-graphitic foams are phenol formaldehyde, resorcinol, formaldehyde, polyurethane, furfural resin, poly vinylidene chloride, poly acrylonitrile, etc. The graphitic foams have been made from coal, pitch, and similar materials [5, 6].

Wang et al. [7] synthesized graphitic foam from coal tar based mesophase pitch at high temperature and pressure. Carbon foam was also synthesized from a bituminous coal with good plasticity properties by thermal process under different temperature and pressure conditions by Calvo et al. [8]. Rios et al. [9] prepared carbon foam by pyrolysis of olive stones under steam.

Replacement of synthetic materials with agricultural products as feedstock in industrial processes has gained momentum all over the world. Carbohydrates can serve as suitable carbon precursors for preparation of carbon foams. Carbohydrates are considered as hydrates of carbon and they easily undergo dehydration to produce carbon. Sucrose is a disaccharide obtained in abundance from sugarcane. Prabhakaran et al. [10] reported synthesis of carbon foam from sucrose. Tondi et al. [11] synthesized carbon foam from low-cost precursors tannin and furfuryl alcohol.

Some other applications of carbon foam are as porous electrodes, high-temperature insulators, filters and demisters, storage batteries, scaffolds and acoustic control medium [12, 13]. One of the promising applications of carbon foam prepared from mesophase pitch is an efficient thermal management material reported by Gallego et al. [14]. The bulk thermal conductivity of this foam was in the range of 40–180 W/m. K. This process does not require the traditional blowing and stabilization steps and therefore, it is cost effective. Initial studies have shown the overall heat transfer coefficients of carbon foam based heat sinks to be up to two orders of magnitude greater than those of conventional heat sinks. Gyenge et al. [15] investigated reticulated vitreous carbon (RVC) electroplated with a Pb–Sn alloy as current collectors for lead-acid batteries. For manufacturing nuclear fuel elements for high-temperature gas-cooled nuclear reactors, RVC could be used for the skeleton material. As for example, Youchison et al. [16] used chemical vapor infiltration technique for thin coating of nuclear fuels (U, Zr, Nb) C on RVC. Karahan et al. [17] studied thermo-mechanical aspect of using a graphitic foam material as the bond between the U–Pu–Zr metallic fuel alloy and the clad for sodium-cooled fast reactors.

This chapter describes the utility of homemade porous carbon (HMPC) as a candidate radionuclide trap material for sodium-cooled fast reactors. The radionuclide-trapping properties of HMPC are also compared with that of the commercially available RVC.

10.2 Need for Radionuclide Trap

10.2.1 Activity Transport

A nuclear reactor is a system in which a controllable nuclear fission chain reaction is maintained. The nuclear reactor is classified according to the velocities of the neutrons, which cause fission as thermal reactors and fast reactors. However, this part of the chapter describes about the activity transport of a sodium-cooled fast reactor and necessity for radionuclide trap.

An operating sodium-cooled fast reactor, in addition to energy, produces radionuclides such as ^{54}Mn, ^{58}Co, ^{60}Co, ^{59}Fe, and ^{51}Cr in the core structural materials by activation due to high neutron flux in the core region. These nuclides are released slowly into sodium coolant due to diffusion and corrosion process. In the event of fuel-clad failure, fission products such as ^{137}Cs, ^{134}Cs, ^{131}I, and ^{140}Ba are released into the coolant stream. These radionuclides are transported to various parts of primary heat transport system and get deposited and subsequently diffused into the structural materials. The phenomenon of release, transport, and deposition of radionuclides in the primary heat transport system of a fast reactor is termed as activity transport. Owing to activity transport, various issues are encountered, which include the following:

(a) high radiation shielding requirement for primary system components and pipelines
(b) restrictions of direct maintenance work
(c) radiological hazards resulting from primary sodium leak
(d) radiation burden during decontamination and disposal of primary components

These issues, all together, describe the radiation burden for operation and maintenance personnel in a fast reactor. It is also essential to keep exposure of plant personnel, from these radiations, to a minimum as stipulated by the ALARA (as low as reasonable achievable) principle [18, 19].

The main sources of radioactive contamination of the primary coolant circuits of a fast reactor are the activated corrosion products from structural materials including fuel-clad, fission products from defective fuel pins and impurities in sodium coolant. In general, the formation of radionuclides takes place in the core of a fast reactor. Thereafter, the radionuclides distribute themselves between coolant, cover gas, and cold trap by processes involving precipitation, adsorption, diffusion, isotopic exchange, and chemical reaction [18–20].

10.2.2 Control of Radionuclide Release

The release of fission product radionuclides in to the primary sodium could be minimized by removing breached fuel immediately upon fission gas detection. However, this conservative fuel management is not acceptable for commercial reactors from economic and plant operation point of view.

For sodium coolant purification, cold trap is used universally. It works on the principle of the reduction of solubility of impurity of interest with decreasing sodium temperature. The principal role of the cold trap is to control the concentration of dissolved oxygen impurity in case of a primary sodium system. The cold trap maintains acceptable levels of activated corrosion products thus minimizing their presence in the other parts of the primary system of a fast reactor.

From the analysis of the cold trap data, it is shown that significant proportion of radioactive inventories can be removed from the coolant, but a desired extent of the removal of activated corrosion products ^{54}Mn, ^{60}Co, and ^{58}Co and fission products ^{137}Cs and ^{134}Cs is unattainable [18]. Therefore, special radionuclide traps are essential. The radionuclide trap is a device located within the reactor primary sodium system that specifically removes activated corrosion products or fission products (in the event of fuel pin failure) from coolant. The material present inside the radionuclide trap device must retain the radionuclides of interest.

As discussed above, the employment of radionuclide trap for the minimization of fission and activated corrosion products in the primary sodium circuits would reduce shielding requirements, radiological hazards resulting from possible primary system leak and simplify direct maintenance work of primary components of a fast reactor. The trap material inside the radionuclide trap device is responsible for binding the radionuclides chemically. This chapter describes the adequacy of carbon foam for trapping ^{137}Cs from liquid sodium.

10.3 Survey of Types of Radionuclide Traps

In order to develop a material to trap cesium, McGuire [21] tested three materials, namely, granular graphite, stainless steel, and nickel as cesium (^{137}Cs) getters in the temperature range 423–823 K in static sodium. The study indicated that graphite is an irreversible getter for ^{137}Cs. It becomes effective at temperature above 473 K and increases its effectiveness rapidly with increasing temperature. Nickel is also effective at 423 K with a gradual release of activity with increasing temperature. Stainless steel gives a response similar to that of nickel.

Cooper et al. [22] measured sorption isotherms for cesium in sodium at various temperatures ranging from 422–811 K using nickel, stainless steel, zirconium and alumina. It was observed that cesium sorption increased with increasing cesium concentration and decreased with temperature. They concluded that nickel and alumina were superior as packing materials for cesium trap.

Clough et al. [23] showed that the volatility of cesium was reduced in presence of graphite due to the formation of cesium–graphite intercalation compounds. Graphite was found to be equally effective for cesium sorption when it was used in the gas phase above liquid sodium. These studies indicated that graphite was a promising material for construction of radionuclide traps for ^{137}Cs and ^{134}Cs. The possibility of graphite acting as a source of carbon, which would carbonize stainless steel elsewhere in the primary circuit, was also considered. It was concluded that certain form of pure carbon was relatively inefficient for carbonization of stainless steel surface when operated at relatively low temperatures.

EBR-II, USA

The graphite and carbon foam, particularly RVC, were exposed to flowing primary sodium in the sodium sampling system at EBR-II [24]. RVC is an open pore, glassified form of amorphous carbon with a high surface-to-volume ratio. The open pore structure of RVC minimizes flow resistance. It showed that both graphite and RVC were effective in trapping ^{137}Cs and ^{134}Cs from primary sodium of EBR-II. RVC was more effective than graphite because of its high surface-to-volume ratio.

The cesium trap designed for operation in the EBR-II primary sodium purification system is shown in Fig. 10.1. Sodium entered the trap through an annular passage, flowed through RVC and exit through the filter. The efficiency for cesium removal per pass through the trap was 90%.

BOR-60 and BR-10, Russia

At BOR-60 [25], two designs of cesium traps, which consisted of a bypass trap and a dummy fuel cartridge, were used as shown in Fig. 10.2. The bypass trap design, as shown in Fig. 10.2a, comprised a cylindrical cartridge containing the layer of specially prepared graphite granules of 5 mm diameter. During 45 h of operation at 463 K, the activity of ^{137}Cs in primary sodium was reduced by a factor of 4 and the gamma dose rate by 2.6. The total activity trapped was 12.6 TBq of ^{137}Cs.

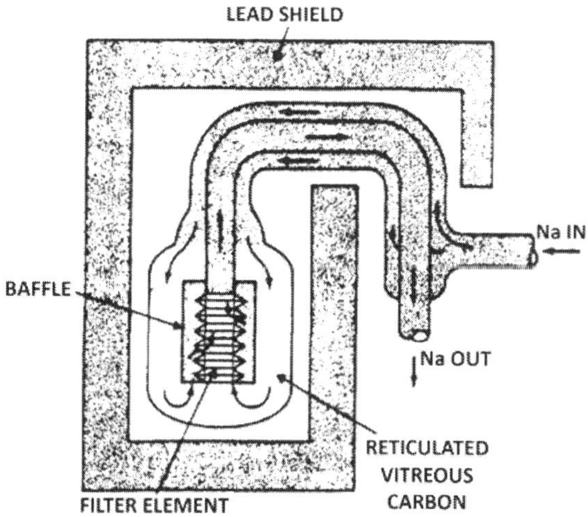

Figure 10.1 EBR-II cesium trap design using RVC [24].

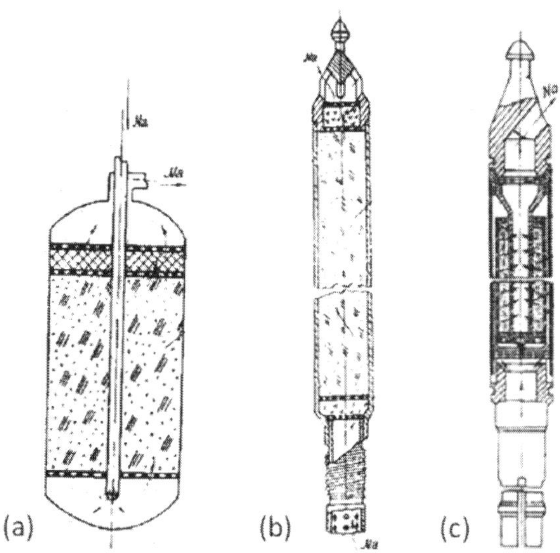

Figure 10.2 Cesium trap design of BOR-60 (a) bypass and (b) dummy fuel cartridge and (c) BR-10 Cesium trap design [25].

To simplify the operation for removal of the trap, a dummy fuel cartridge was designed as shown in Fig. 10.2b. The shape of the trap was similar to that of a fuel subassembly. Decontamination factors of 4.2 and 3.5 in activity and 2.7 and 1.4 in dose rate were obtained during two consecutive tests. In these two tests, 8 and 2.2 TBq of ^{137}Cs was removed from the primary circuits.

At BR-10 [25], the cesium traps were installed in a central channel instead of an experimental fuel subassembly. The head and tail parts of standard fuel subassembly were used in this design as shown in Fig. 10.2c. The tests were carried out before the reactor shutdown. The traps with pyrographite and activated charcoal were loaded alternatively into the central channel. As a result of purification, the gamma dose rate on the pipe surface was reduced by a factor of 3. The cesium (^{137}Cs and ^{134}Cs) concentration was reduced by a factor of 18.

KNK-II, Germany

In KNK-II, three materials, namely RVC, SIGRAFLEX (flexible foil graphite), and SIGRATEX (textile tissues), were exposed to the flowing coolant in the primary sodium sampling station [26]. Four experimental cesium traps were built, and two of these were filled with RVC, one with SIGRATEX and other with SIGRAFLEX. During the trap experiment, the ^{137}Cs activity in the KNK-II primary sodium was 0.25 kBq/g sodium. In 48 h of accumulated operation, the traps removed 2150 MBq ^{137}Cs and 55 MBq ^{134}Cs from the primary sodium.

FFTF, USA

At the Fast Flux Test Facility (FFTF), the cesium trap was installed to remove radioactive cesium from the primary sodium circuits. The FFTF cesium trap was made by modifying spare cold trap components [27]. The trap assembly consisted of an economizer, collector tank and a coolant jacket. The collector tank contained discs of RVC inside the flow divider. The temperature of the trap was maintained above 450 K to prevent precipitation of oxides and hydrides in the trap. The RVC was retained by a sintered metal filter of 25 μm pore size. This is to prevent particle release from the cesium trap into the sodium circuits. The dissolution rate of carbon in sodium was monitored and it was found that less than 10 g carbon would be dissolved during 100 days of operation of the trap (~2.3 kg) at 505 K, which was acceptable. A flow rate of

22.7 L/min was maintained and this caused pressure drop of 35 kPa in the trap. This trap brought down the ^{137}Cs activity from 6290 Bq/g to 2442 Bq/g. The trap removed 60% of the ^{137}Cs inventory during 33 days of operation.

RAPSODIE, France

For purification of RAPSODIE [28] primary sodium during reactor operation, four traps were designed in the shape of subassembly consisting of hexagonal prism (410 mm in length). In the inner cylindrical housing of the trap, eight blocks of RVC, each 50 mm in length and 23 mm in diameter were superimposed. Gamma spectrometry analysis showed that RVC extracted 490 GBq of ^{137}Cs from primary sodium.

JOYO, Japan

In JOYO, the cesium trap using RVC was installed in the primary coolant sodium purification system to capture radioactive cesium released from failed fuels. The cesium loading capacity of trap was designed to be 7.4×10^{12} Bq (200 Ci). Since the reactor did not experience any fuel pin failure, fission products contamination was not an issue in this reactor [29].

BN-350, Kazakhstan

Seven cesium traps employing RVC were fabricated and used to remove ^{137}Cs from primary sodium of BN-350 [30]. The trap design was based on the out-of-core design similar to that of EBR-II. Each cesium trap continued its operation until the RVC became saturated with cesium, after which it was removed and replaced by a new trap. The trap experiments indicated that 255.3 TBq of ^{137}Cs was trapped by RVC, which represented 98.6% of estimated initial ^{137}Cs inventory in the primary system. The specific activity of ^{137}Cs in BN-350 primary sodium was decreased by a factor of 800 (296 MBq/kg to 0.33 MBq/kg) by 16 kg of RVC.

FBTR, India

At FBTR, cesium trap employing RVC in the form of fuel subassembly was fabricated for incorporation in reactor core [20]. This trap is for operation at 473 K during shutdown period in case of fuel pin failure.

10.4 Experimental Methods

10.4.1 Synthesis of Carbon Foam

The carbon foam HMPC was synthesized from sucrose. An acidic solution containing 30 g of sucrose in 100 mL of distilled water with 1 mL of nitric acid was prepared in a 500 mL capacity Teflon beaker. The pH of this solution was 0.85. The solution was concentrated by heating at 383 K on a hot plate with constant stirring (100 rpm) till it turned into a dark viscous resin. After a while, the resin started foaming and the soft foam was heated in an air oven at 383 K for a period of 24 h for complete foaming and drying. The green foam thus obtained was cut into regular shapes and sintered under an argon atmosphere with a heating rate of 100 K/h and 2 h dwell time. Fig. 10.3 shows the flowchart for preparation of HMPC.

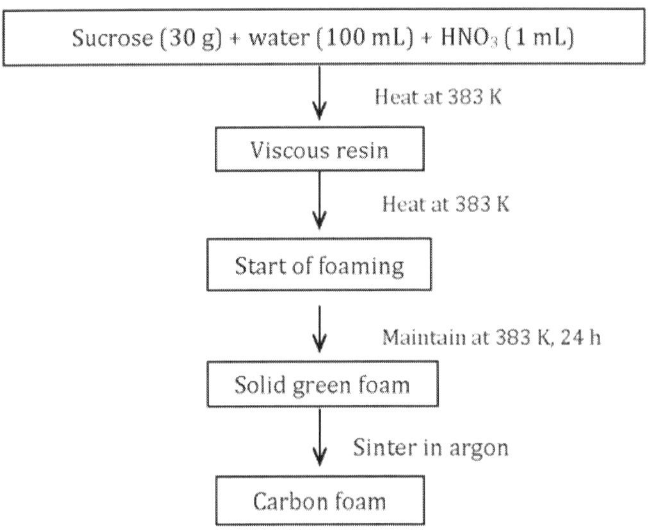

Figure 10.3 Flowchart for preparation of carbon foam.

10.4.2 Characterization

The green foam and sintered foam materials were characterized with a variety of analytical methods. For the purpose of thermal analysis, X-ray diffraction and infrared analysis, the green foam and sintered foam were crushed to a fine powder with a mortar and pestle. The pore size of the foam samples was calculated from scanning electron microscopic examination. Details of techniques used for characterization of foam samples are described below.

The thermal stability of the green foam was studied under argon atmosphere at a constant heating rate of 10 K/min up to 1273 K. A high-temperature thermogravimetric analyzer (TGA, SETSYS 16/18 Evolution, Setaram, France) with a resolution of 1 μg was employed. For the purpose of evolved gas analysis, a quadrupole mass spectrometer (Pfeiffer Vacuum, Germany) was employed. The green foam in fine powder form was loaded in an alumina crucible positioned in the microbalance of TGA. After loading the sample, the TGA system was evacuated and filled with argon gas. Evacuation and argon filling was done twice to minimize the residual moisture/air levels in the furnace atmosphere. Argon was passed over the sample at a rate of 20 mL/min throughout the experiment.

The X-ray diffraction technique was employed for the phase analysis of the foam materials at various steps of synthesis. The room temperature X-ray diffractograms of HMPC (green foam and foams sintered at high temperature), RVC and graphite were recorded using X-ray diffractometer (Siemens D500, Germany) with Cu $K\alpha$ radiation (λ = 0.15406 nm). The X-ray diffraction patterns were recorded in the angular range of 10–80° with a step size of 0.05°/sec.

Infrared (IR) spectra of HMPC (green foam and foams sintered at high temperature) were recorded using the Fourier transform infrared (FTIR) spectrometer (BOMEM MB100, USA) in the frequency range 4000–400/cm with a resolution of 4/cm by the pressed KBr pellet technique. A ratio of 1:100 was maintained between the sample and KBr.

The surface morphology of the foams was examined by scanning electron microscope (SEM). In the SEM, high-energy electrons interact with the specimen and the collected secondary and backscattered electrons are utilized for imaging. This technique demands that the sample to be analyzed be electrically conducting. The micrographs of conducting samples were obtained as such. For non-conducting and low conducting samples, the surface was coated with gold-palladium alloy prior to SEM examination.

The micrographs of green and sintered foams of HMPC as well as that of RVC were obtained using SEM (Philips, XL30, The Netherlands). Since the green foam was electrically non-conducting, they were coated with gold–palladium alloy of 15–20 nm thicknesses using a sputter coater (VG Microtech, UK) prior to SEM examination.

The weight of the foam samples was measured by microbalance. Since HMPC could be shaped, the volume was readily calculated by dimension measurements and therefore, density of foam materials were obtained by weight and volume ratio. The porosity of HMPC was calculated using the following equation given below [31, 32]:

$$\text{Porosity} = 100(1 - \rho_b/\rho_t),$$

where ρ_t is the theoretical density and ρ_b is the bulk density of the material.

10.4.3 Cesium Trap Studies

The cesium trap experiments were conducted in a static sodium system. Figure 10.4 shows the schematic of the experimental setup. The experimental vessel was made of stainless steel with knife-edged flanges at the top to ensure hermetic sealing. The vessel has a side port for providing argon cover gas above the sodium surface. The sample holder could be positioned anywhere within the experimental vessel by means of a Veeco-type sliding seal. Vapor suppressors were provided to prevent loss of sodium by evaporation during the experiment. HMPC (sintered at 1223 K) and commercial RVC in the cylindrical form were used as samples. The uptake of ^{137}Cs by HMPC and RVC was monitored using a NaI(Tl) detector system.

The foam specimens were preconditioned by heating at 473 K under vacuum to ensure the filling of sodium into the pores while immersing into liquid sodium. This process is called sodium wetting. Experimental vessel for wetting the carbon foam specimens with sodium was assembled in inert atmosphere glove box. The sample holder containing foam specimens was positioned above the sodium surface at 473 K. Prior to immersion of the foam sample in liquid sodium, the vessel was evacuated. The foam was immersed in sodium under vacuum and subsequently filled with argon to facilitate impregnation of the pores by liquid sodium.

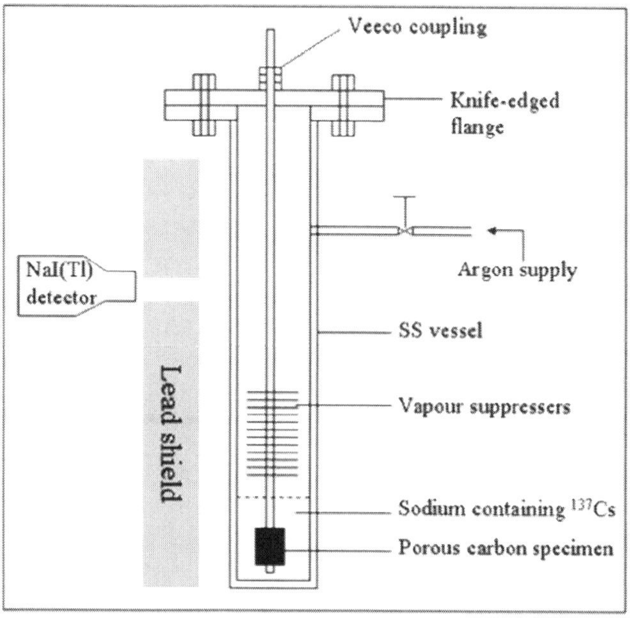

Figure 10.4 Schematic of experimental setup for ^{137}Cs trap studies.

The cesium (^{137}Cs) trap experimental setup was handled in inert atmosphere glove box. The known quantity of ^{137}Cs was transferred to liquid sodium by the planceting technique. Around 23 g of sodium were taken in the vessel and heated to 473 K for 24 h. The sample holder with the preconditioned carbon foam sample was dipped in sodium containing ^{137}Cs. The foam specimen was withdrawn periodically from liquid sodium to the counting position and the trapped activity in the samples was counted by NaI(Tl) detector. For each sample, three experimental runs were carried out at 473 K.

10.4.4 Radioactivity Counting

Cesium activities (^{137}Cs) in HMPC and RVC samples were measured by standard gamma counting technique using a NaI(Tl) scintillation detector with single channel analyzer at periodic intervals during the experiment. For this purpose, PMT Nucounter attached with NaI(Tl)/PMT-based detector (Para Electronics Pvt Ltd., Mumbai, India) was employed.

10.5 Results and Discussion

For the synthesis of carbon foam mentioned in this chapter, sucrose was used, which played an important role to form a porous network. Sucrose is a carbohydrate containing a six-member ring of glucose and a five-member ring of fructose. It is known that sucrose in the acidic aqueous medium undergoes hydrolysis to form glucose and fructose as shown below.

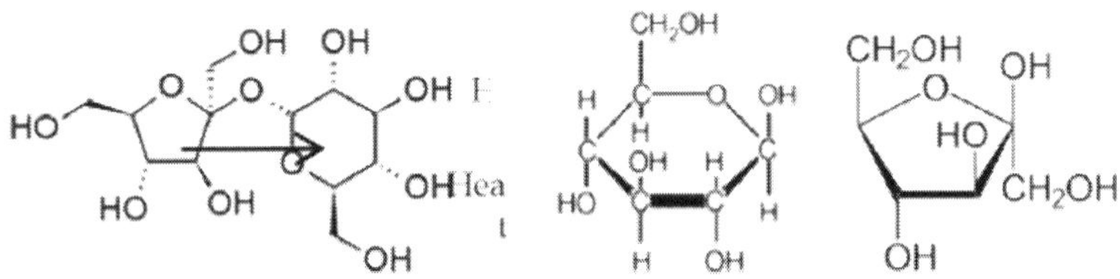

Sucrose Glucose Fructose

Acid (in the present case, nitric acid) further oxidizes the glucose and fructose to a mixture of hydroxy carboxylic acids such as gluconic acid, saccharic acid, glycolic acid, and tartaric acid [33]. Upon concentrating by heating, the nitric acid–sucrose solution undergoes polymerization followed by condensation to form a resin. After several hours, the resin forms a solid green foam [34].

For the synthesis of carbon foam termed as HMPC, sucrose acts as a base material and nitric acid as the pore former. Upon heating, the aqueous solution of sucrose-HNO_3 underwent polymerization and condensation to form a resin. The resin subsequently started foaming. This was confirmed by the fact that sucrose did not undergo foaming when heated at 383 K in the absence of HNO_3. The green foam was heated to high temperature to obtain sintered HMPC. Figure 10.5 shows photograph of (a) HMPC (green foam) and (b) HMPC sintered at 1223 K. Since HMPC could be shaped, the bulk density was readily calculated from their weight and volume. The calculated bulk density of HMPC is 0.10 g/cc and porosity is 95.5%. The calculated density and porosity of RVC is 0.05 g/cc and 98%, respectively, while the reported porosity of RVC was 97% [12]. Thus an ultra light product with extremely high porous HMPC was prepared.

Figure 10.5 Photograph of (a) HMPC (green foam) and (b) HMPC sintered at 1223 K.

The mass loss and gases evolved from the green foam during sintering were studied using the TGA coupled with EGA using argon as carrier gas. The typical results of thermal decomposition of HMPC (green foam) at a constant heating rate of 10 K/min are shown in Fig. 10.6. It showed a three-stage decomposition of green foam (stage I: 310–430 K, stage II: 430–610 K, and stage III: 610–700 K) with 70% mass loss. Stage I was due to removal of H_2O from the green foam. Stage II and stage III were due to the decomposition of volatile organic compounds present in the green foam.

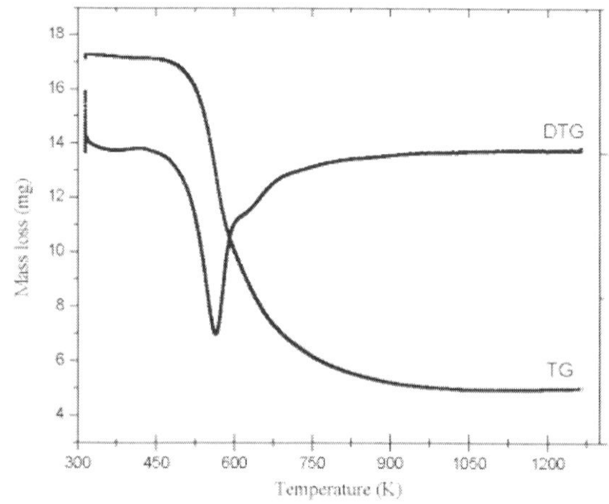

Figure 10.6 Thermal decomposition pattern of HMPC (green foam).

The decomposition products of HMPC were analyzed using EGA as a function of temperature. The gases evolved from green foam are presented in Fig. 10.7. It was observed that gaseous species with mass numbers 2 (H_2), 18 (H_2O), 28 (CO), 30 (NO), 32 (O_2), 44 (CO_2), and 46 (NO_2) were evolved from the green foam.

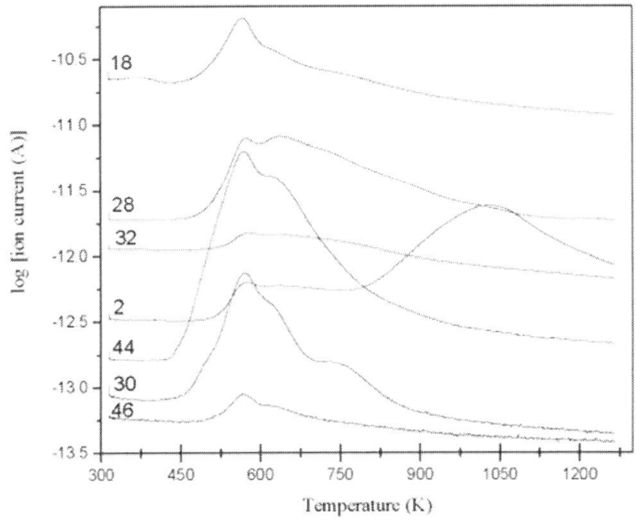

Figure 10.7 Concentrations of evolved gases from HMPC (green foam) as measured from the respective ion currents as a function of temperature.

The IR absorption spectra of HMPC, both green foam and that sintered at high temperature are shown in Fig. 10.8. The green foam showed five absorption peaks at 3398, 2926, 1726, 1641 and 1035/cm as shown in Fig. 10.8a. These peaks were attributed to –O–H, –C–H, –O–C=O, –C=C–, and –C–O–C (stretches), respectively. But the peaks at 3398, 2926, and 1035/cm were not present in the foam sintered at 573 K as well as that sintered at 773 K as shown in Figs. 10.8b,c. This was due to the removal of –O–H (as H_2O) and breaking of –C–H and –C–O–C bonds in the green foam at the temperature. However, the foam sintered at 973 K and above did not exhibit any characteristic IR feature as shown in Figs. 10.8d,e, thus indicating that the carbon foam sintered at 973 K and above was free of organic compounds.

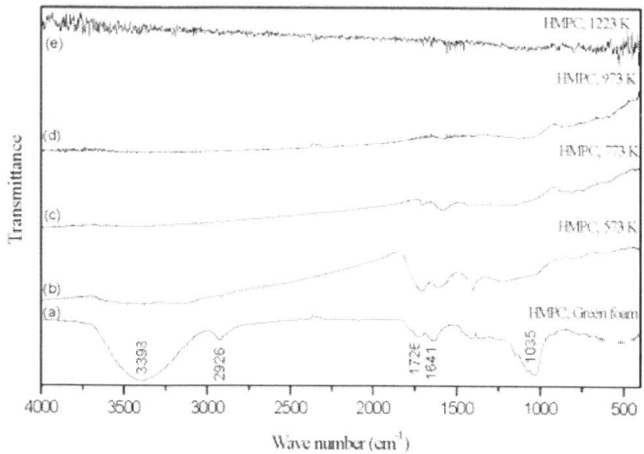

Figure 10.8 FTIR spectra of HMPC (green foam and foams sintered at different temperatures).

Figure 10.9 shows the XRD patterns of HMPC (green foam and foams sintered at higher temperature), RVC and graphite. HMPC (green foam) showed broad peak at 2θ value 19 indicating amorphous nature. HMPC sintered at 973 and 1223 K as well as that of RVC and graphite showed two peaks at 2θ values of (22.2, 43.5), (23, 44), (21.6, 43.6) and (26.3, 44.4), respectively. The low-intensity peak at 2θ value of nearly 44 was found to appear after sintering the foam at 973 K and above. This was attributed to the slow graphitization of the foam at high temperatures employed. These values could be compared with two intense peaks at 2θ values of 26.55 and 44.57 corresponding to the planes (002), (101) (JCPDS file No. 89-8487) indicating hexagonal primitive crystal structure of graphite. The average inter planar distance (d-spacing between two graphitic planes) has been calculated using Bragg's equation ($n\lambda = 2d \sin\theta$) and the values are 0.40, 0.38, 0.41, and 0.33 nm for HMPC sintered at 973 K, 1223 K, RVC, and graphite, respectively. Table 10.1 shows the values of 2θ, full width half maxima (FWHM) and d-spacing of HMPC, RVC and graphite. It is observed that 2θ values increase, while values of FWHM decrease with increase in sintering temperature, indicating improvement in crystallinity of HMPC during sintering. It can also been seen that the crystallinity of HMPC is higher than that of RVC.

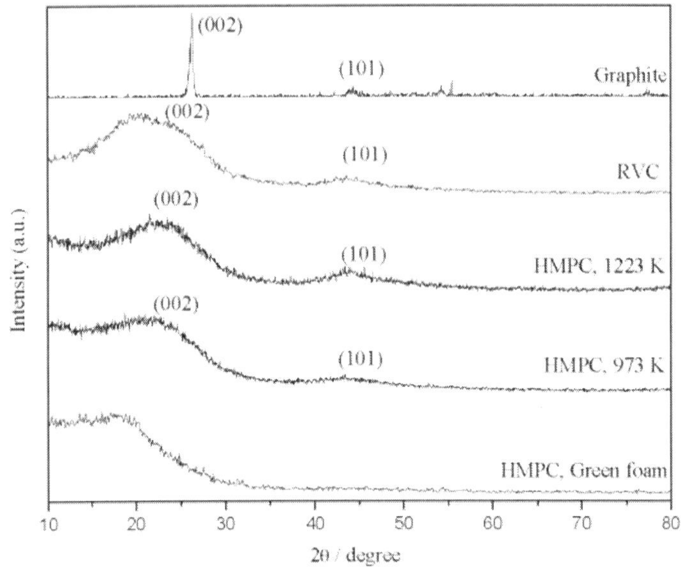

Figure 10.9 XRD patterns of HMPC (green foam and foams sintered at high temperatures). RVC and graphite patterns are given for comparison.

Table 10.1 Parameters obtained from XRD measurements of HMPC, RVC, and graphite

Sample	2θ (degree)	FWHM (degree)	d-spacing (nm)
HMPC sintered at 973 K	22.2	9.5	0.40
HMPC sintered at 1223 K	23	9.0	0.38
RVC	21.6	11.75	0.41
Graphite	26.3	0.4	0.33

Figure 10.10 shows the secondary electron images of green foam, foams sintered at higher temperatures and RVC. The green foam did not exhibit extensive porous network, but the HMPC sintered at 973 K and above exhibited regular porous network. From the images, the average pore size was determined to be 450

and 550 μm for HMPC sintered at 973 K and 1223 K, respectively. The average pore size of RVC is 600 μm and it is fully open porous network as shown in Fig. 10.10d. HMPC exhibited pores of near spherical configuration that were interconnected by other pores. The edges of the pores were more irregular and bold compared to that observed in RVC. This is attributed to the difference in the methods of preparation. HMPC was synthesized from sucrose and presumably RVC was synthesized by template method.

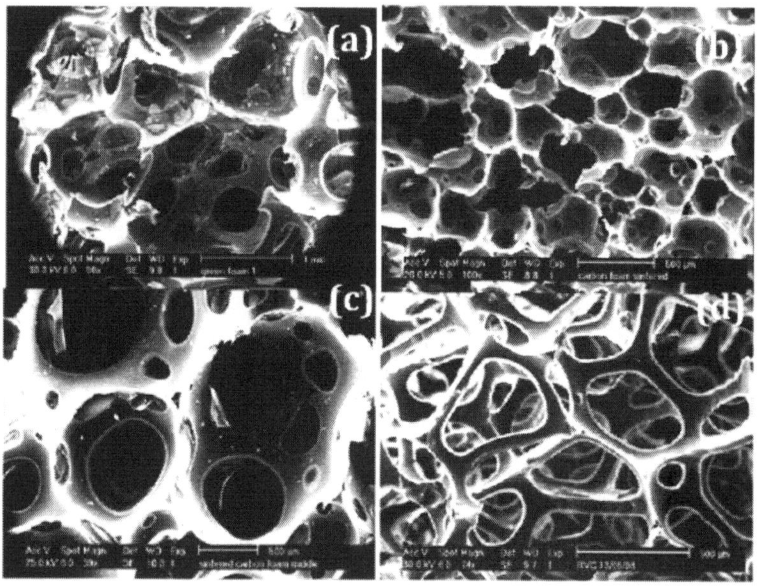

Figure 10.10 Secondary electron image of (a) HMPC (green foam), (b) HMPC sintered at 973 K, (c) HMPC sintered at 1223 K and (d) RVC.

Figure 10.11 shows secondary electron images of powder form of HMPC. Particles were irregular in size of ~50 μm. HMPC sintered at 973 K exhibited presence of carbon spheres as shown in Fig. 10.12. The carbon spheres, however, were not observed in other sintered HMPC specimens. Mi et al. [35] synthesized carbon microspheres by glucose (carbohydrate) hydrothermal method. This indicates that carbon microspheres can be synthesized from sucrose by modifying the synthesis procedure of carbon foam.

Figure 10.11 Secondary electron image of powder form of (a) HMPC sintered at 973 K and (b) HMPC sintered at 1223 K.

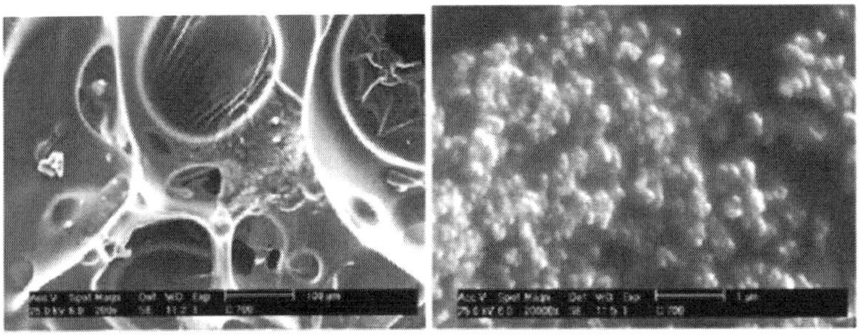

Figure 10.12 Secondary electron image of carbon spheres on HMPC sintered at 973 K.

10.6 Cesium Trap Studies

The cesium (^{137}Cs) trap experiments were conducted in a static sodium system with HMPC in cylindrical form as samples. Three experiments, namely Run 1, Run 2, and Run 3, were carried out at 473 K. The ^{137}Cs activities taken in sodium for these runs were 15, 13, and 16 μCi, respectively. The ^{137}Cs activities trapped by HMPC from sodium were 11, 10, and 12 μCi for the respective runs. The ^{137}Cs uptake behavior of HMPC is shown in Fig. 10.13. In all the runs, the uptake of ^{137}Cs by HMPC was found to saturate after 200 h. The trapping efficiencies of HMPC for ^{137}Cs were calculated and found to be 73, 77, and 75% for Run 1, Run 2, and Run 3, respectively [34]. The distribution coefficients (K_D) of ^{137}Cs between HMPC and sodium were computed using the following equation:

$$K_D = \left(\frac{\text{No. of moles of cesium in carbon}}{\text{No. of moles of carbon}}\right) \Big/ \left(\frac{\text{No. of moles of cesium in sodium}}{\text{No. of moles of sodium}}\right)$$

The K_D values of ^{137}Cs between HMPC and sodium were found to be in the range of $4.9–6.0 \times 10^2$.

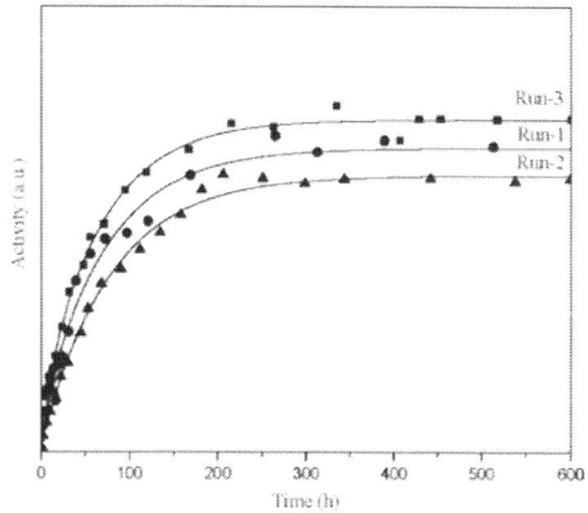

Figure 10.13 Cesium (^{137}Cs) uptake behavior of HMPC.

For comparison with HMPC, the trap experiments, using commercially available RVC, were also carried out at 473 K. Three experiments, namely Run 4, Run 5, and Run 6, were carried out. The ^{137}Cs activities taken for these trap experiments were 14, 30, and 12 μCi, respectively. The activities trapped by RVC were 12.4, 28, and 10 μCi for the respective runs. The uptake of ^{137}Cs by RVC was also found to saturate after 200 h as shown in Fig. 10.14. The trapping efficiencies of RVC for ^{137}Cs were in the range of 83–93% [34]. The K_D values of ^{137}Cs between RVC and sodium were in the range of $1.4–3.5 \times 10^3$.

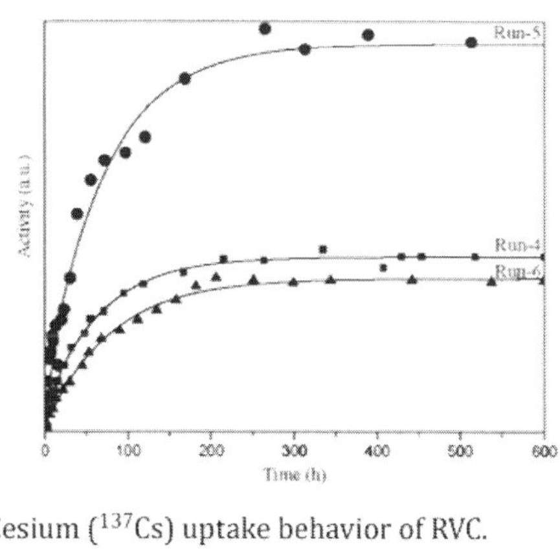

Figure 10.14 Cesium (^{137}Cs) uptake behavior of RVC.

Table 10.2 shows the relative efficiency of HMPC and RVC for trapping ^{137}Cs from sodium and the K_D values. The relatively lower trapping efficiency and smaller distribution coefficient associated with HMPC compared to that of RVC are attributed to porosity and density differences between them. However, it is to be noted that the trapping efficiency of HMPC for ^{137}Cs is adequate enough to be a getter material for a radionuclide (^{137}Cs) trap device.

Table 10.2 Details of ^{137}Cs trap experiment employing (a) HMPC and (b) RVC

	Cesium activity (μCi)			Wt. of trapping material (g)	Wt. of sodium (g)	K_D	Efficiency (%)
	Taken	Trapped	Remaining				
HMPC							
Run 1	15	11	4	0.066	23	4.9×10^2	73
Run 2	13	10	3	0.067	23	6.0×10^2	77
Run 3	16	12	4	0.072	23	4.9×10^2	75
RVC							
Run 4	14	12.4	1.6	0.052	22	1.7×10^3	88
Run 5	30	28	2	0.047	23	3.5×10^3	93
Run 6	12	10	2	0.042	23	1.4×10^3	83

10.7 Summary and Conclusions

HMPC was synthesized from sucrose and characterized by various techniques. The calculated bulk density of HMPC was 0.10 g/cc and the porosity was 95.5%. From the XRD measurements, it was observed that 2θ values increase while values of FWHM decrease with increase in sintering temperature indicating improvement in crystallinity of HMPC. The average pore size was determined to be 450 and 550 μm for HMPC sintered at 973 and 1223 K, respectively. HMPC exhibited pores of near spherical configuration that are interconnected by other pores. HMPC sintered at 1223 K was employed as specimen for cesium (^{137}Cs) trap study.

In the trap experiment at 473 K using HMPC, the ^{137}Cs activities taken were in the range of 13–16 μCi and that trapped by

HMPC were in the range of 10–12 μCi. The trapping efficiencies of HMPC for ^{137}Cs were in the range of 73–77%. The values of distribution coefficient (K_D) of ^{137}Cs between HMPC and sodium were found to be in the range of 4.9–6.0 × 10^2. HMPC exhibited saturation for ^{137}Cs uptake after 200 h.

For comparison with HMPC, the trap experiment using commercially available RVC was also carried out at 473 K. The ^{137}Cs activities taken for the trapping experiments were in the range of 12–30 μCi and activities trapped by RVC were in the range of 10–28 μCi. RVC was also found to saturate for ^{137}Cs uptake after 200 h. Trapping efficiencies of RVC for ^{137}Cs were in the range 83–93%. The K_D values of ^{137}Cs between RVC and sodium were in the range of 1.4–3.5 × 10^3.

The relatively lower trapping efficiency and smaller distribution coefficient associated with HMPC compared to that of RVC were attributed to porosity and density differences between HMPC and RVC. However, it is to be noted that the trapping efficiency of HMPC for ^{137}Cs is adequate enough to be a getter material for radionuclide (^{137}Cs) trap device.

Acknowledgments

The authors acknowledge Mr. P. Muralidaran, Ms. R. Sudha, Mr. A. Manivannan, Mr. Swapan Kumar Mahato, Mr. Raja Madhavan, Mr. N. Ramanathan, and Mr. Sajal Ghose for their support. The authors also acknowledge Dr. K. S. Viswanathan, Dr. R. Viswanathan, and Dr. P. R. Vasudeva Rao for constant encouragement during the course of this work.

References

1. Ford, W. (1964). Method of making cellular refractory thermal insulating material, US patent 3121050.
2. Googin, J. M., Napier, J. M., and Scrivner, M. E. (1967). Method for manufacturing foam carbon products, US patent 3345440.
3. Klett, R. D. (1975). High temperature insulating carbonaceous material, US patent 3914392.
4. Vinton, C. S., and Franklin, C. H. (1977). Method for the preparation of vitreous carbon, US patent 4022875.

5. Fujii, M., and Minohata, M. (1990). Method of producing elastic graphite structure, US patent 4908200.
6. Min, G., Zengmin, S., Weidong, C., and Hui, L. (2007). Anisotropy of mesophase pitch-derived carbon foams, *Carbon*, **45**, pp. 141–145.
7. Wang, X., Zhong, J., Wang, Y., Yu, M., and Wang, Y. (2007). The study on the formation of graphite foam, *Mater. Lett.*, **61**, pp. 741–746.
8. Calvo, M., Garcia, R., Arenillas, A., Suarez, I., and Moinelo, S. R. (2005). Carbon foams from coals. A preliminary study, *Fuel*, **84**, pp. 2184–2189.
9. Rios, R. V. R. A., Martinez-Escandell, M., Molina-Sabio, M., and Rodriguez-Reinoso, F. (2006). Carbon foam prepared by pyrolysis of olive stones under steam, *Carbon*, **44**, pp. 1448–1454.
10. Prabhakaran, K., Singh, P. K., Gokhale, N. M., and Sharma, S. C. (2007). Processing of sucrose to low-density carbon foams, *J. Mater. Sci.*, **42**, pp. 3894–900.
11. Tondi, G., Fierro, V., Pizzi, A., and Celzard, A. (2009). Tannin-based carbon foams, *Carbon*, **47**, pp. 1480–1492.
12. Product note on reticulated vitreous carbon, Energy Research and Generation Incorporation, Oakland, California.
13. Wang, J. (1981). Reticulated vitreous carbon: A new versatile electrode material, *Electrochim. Acta*, **26**, pp. 1721–1726.
14. Gallego, N. C., and Klett, J. W. (2003). Carbon foams for thermal management, *Carbon*, **41**, pp. 1461–1466.
15. Gyenge, E., Jung, J., and Mahato, B. (2003). Electroplated reticulated vitreous carbon current collectors for lead-acid batteries: Opportunities and challenges, *J. Power Sources*, **113**, pp. 388–395.
16. Youchison, D. L. (2010). Methods for manufacturing porous nuclear fuel elements for high-temperature gas-cooled nuclear reactors, US Patent 7666463 B1.
17. Karahan, A., and Kazimi, M. S. (2010). Using graphite foam as the bonding material in metal fuel pins for sodium fast reactors, *Nuclear Fuel and Structural Material Embedded Topical Meeting, American Nuclear Society Annual Meeting*, pp. 790–792.
18. Feuerstein, H., Hooper, A. J., and Johnson, F. A. (1979). Mechanism of release of radioactive products into liquid-metal coolants, their transport within the circuits and removal from LMFBRs, *Atomic Eng. Rev.*, **17**, pp. 697–761.

19. Technical document on "Fission and corrosion product behavior in liquid metal fast breeder reactors (LMFBRs)", (1993). *IAEA-TECDOC-***687**, pp. 9–155.

20. Ganesan, V., and Muralidaran, P. (2007). Activity transport in fast reactor heat transport systems, *Proceedings of the Nuclear Radiochemistry Symposium (NUCAR-2007)*, Vadodara, pp. 69–76.

21. McGuire, J. C. (1966). Removal of ^{137}Cs from molten sodium by means of getters, *Trans. Am. Nucl. Soc.*, **9**, pp. 303–304.

22. Cooper, M. H., and Taylor, G. R. (1971). Measurements of cesium isotherms in liquid sodium, *Nucl. Technol.*, **12**, pp. 83–92.

23. Clough, W. S., and Wade, S. W. (1971). Caesium behaviour in liquid sodium: The effect of carbon, *J. Nucl. Eng.*, **25**, pp. 445–456.

24. Olson, W. H., and Ruther, W. E. (1979). Controlling cesium in the coolant of the experimental breeder reactor II, *Nucl. Technol.*, **46**, pp. 318–322.

25. Krasnoyarov, N. V., Polyakov, V. I., Sobolev, A. M., and Yakshin, E. K. (1984). The removal of cesium from the fast reactor primary sodium coolant, *Proceedings of the 3rd International Conference on Liquid Metal Engineering and Technology (LIMET-1984)*, Oxford, **3**, pp. 185–189.

26. Stade, K. C., Stamm, H. H., Hanke, H. D., and Clauss, H. (1985). Status of cesium removal from primary sodium in the German LMFBR program, *Proceedings of the International Topical Meeting on Fast Reactor Safety*, Knoxville, Tennessee, pp. 197–203.

27. Bechtold, R. A., and Grenard, C. E. (1988). FFTF cesium trap design, installation and operating experience, *Proceedings of the 4th International Conference on Liquid Metal Engineering and Technology (LIMET-1988)*, Avignon, France, ED SFEF F-75724, pp. 609/1–10.

28. Hanebeck, N., Msika, D., Misraki, J., Allegre, R., Tusche, R., and Stade, K. C. (1988). Development of cesium traps for commercial sodium-cooled fast breeder reactors: ELCESNA, KNK II, RAPSODIE, *Proceedings of the 4th International Conference on Liquid Metal Engineering and Technology (LIMET-1988)*, Avignon, France, ED SFEF F-75724, pp. 631/1–14.

29. Aoyama, T., Odo, T., Suzuki, S., and Yogo, S. (2002). Operational experience and upgrading program of the experimental fast reactor JOYO, *IAEA-TECDOC-1405*, pp. 29–61.

30. Romanenko, O. G., Allen, K. J., Wachs, D. M., Planchon, H. P., Wells, P. B., Nazarenko, P., Dumchev, I., Maev, V., Zemtzev, B., Tikhomirov, L.,

Yakovlev, V., and Synkov, A. (2005). Cleaning cesium radionuclides from BN-350 primary sodium, *Nucl. Technol.*, **150**, pp. 79–99.

31. Han, Y. S., Li, J. B., and Chem, Y. J. (2003). Fabrication of porous bimodal alumina ceramics, *Mater. Res. Bull.*, **38**, pp. 373–379.
32. Prabhakaran, K., Gokhale, N. M., Sharma, S. C., and Lal, R. (2005). A novel process for low-density alumina foams, *J. Am. Ceram. Soc.*, **88**, pp. 2600–2603.
33. Finar, I. L. (1973). *Organic Chemistry vol. I, The Fundamental Principle*, 6th ed., ELBS, London, pp. 503–530.
34. Jana, P., and Ganesan, V. (2009). Synthesis, characterization and radionuclide (^{137}Cs) trapping properties of a carbon foam, *Carbon*, **47**, pp. 3001–3009.
35. Mi, Y., Hu, W., Dan, Y., and Liu, Y. (2008). Synthesis of carbon microspheres by a glucose hydrothermal method, *Mater. Lett.*, **62**, pp. 1194–1196.

Index